CAMBRIDGE STUDIES IN
ADVANCED MATHEMATICS 9

Characteristic classes and the cohomology of finite groups

Characteristic classes and the cohomology of finite groups

C. B. THOMAS
Department of Pure Mathematics and Mathematical Statistics, University of Cambridge

The right of the University of Cambridge to print and sell all manner of books was granted by Henry VIII in 1534. The University has printed and published continuously since 1584.

CAMBRIDGE UNIVERSITY PRESS
CAMBRIDGE
LONDON NEW YORK NEW ROCHELLE
MELBOURNE SYDNEY

CAMBRIDGE UNIVERSITY PRESS
Cambridge, New York, Melbourne, Madrid, Cape Town, Singapore, São Paulo, Delhi

Cambridge University Press
The Edinburgh Building, Cambridge CB2 8RU, UK

Published in the United States of America by Cambridge University Press, New York

www.cambridge.org
Information on this title: www.cambridge.org/9780521256612

First published 1986
This digitally printed version 2008

A catalogue record for this publication is available from the British Library

Library of Congress Cataloguing in Publication data
Thomas, C.B. (Charles Benedict)
Characteristic classes and the cohomology of finite groups.
(Cambridge studies in advanced mathematics; 9)
1. Finite groups. 2. Homology theory.
3. Characteristic classes. I. Title. II. Series.
QA171.T48 1986 512′.2 85–17442

ISBN 978-0-521-25661-2 hardback
ISBN 978-0-521-09065-0 paperback

Characteristic classes and the cohomology of finite groups

C. B. THOMAS
Department of Pure Mathematics and Mathematical Statistics,
University of Cambridge

CAMBRIDGE UNIVERSITY PRESS

CAMBRIDGE
LONDON NEW YORK NEW ROCHELLE
MELBOURNE SYDNEY

CAMBRIDGE UNIVERSITY PRESS
Cambridge, New York, Melbourne, Madrid, Cape Town, Singapore, São Paulo, Delhi

Cambridge University Press
The Edinburgh Building, Cambridge CB2 8RU, UK

Published in the United States of America by Cambridge University Press, New York

www.cambridge.org
Information on this title: www.cambridge.org/9780521256612

First published 1986
This digitally printed version 2008

A catalogue record for this publication is available from the British Library

Library of Congress Cataloguing in Publication data
Thomas, C.B. (Charles Benedict)
Characteristic classes and the cohomology of finite groups.
(Cambridge studies in advanced mathematics; 9)
1. Finite groups. 2. Homology theory.
3. Characteristic classes. I. Title. II. Series.
QA171.T48 1986 512′.2 85–17442

ISBN 978-0-521-25661-2 hardback
ISBN 978-0-521-09065-0 paperback

For Maria

Contents

Introduction

If G is an arbitrary finite group (more generally a finitely presented possibly infinite group) it is an easy exercise in combinatorial topology to construct a finite 2-dimensional simplicial complex with fundamental group isomorphic to G. By attaching simplexes of dimension $\geqslant 3$ in a systematic way it is possible to embed this 2-complex in a larger complex $K(G, 1)$ without changing the fundamental group, but choosing K to have universal covering space homotopy equivalent to a point. One may then define the cohomology groups $H^k(G, \mathbb{Z})$ of the discrete group G to be the cohomology groups of the space K. This definition is independent of the topological model chosen, and may indeed be copied algebraically with the coefficients $\mathbb{Z}$ replaced by some (left) $\mathbb{Z}G$-module A. It is clear that the graded ring $\{H^k(G, \mathbb{Z}), k \geqslant 0\}$ is an important invariant of the group, indeed if the homomorphism $\varphi: G_1 \to G_2$ of finite groups induces an isomorphism $\varphi^*: H^*(G_2, \mathbb{Z}) \to H^*(G_1, \mathbb{Z})$, then the groups G_1 and G_2 are isomorphic. However although the literature contains one or two striking applications of group cohomology, for example to the construction of infinite class field towers and to the study of outer automorphisms of p-groups, its systematic use as a tool has been held up by a lack of calculations for specific groups. The aim of this book is to remedy this situation partially, by exploring the connection between complex representations and integral cohomology provided by characteristic classes. In this way we obtain a subring $\mathrm{Ch}(G)$ of $H^{\text{even}}(G, \mathbb{Z})$, over which, as a consequence of the Hilbert basis theorem, the integral cohomology is finitely generated as a module. Warning: if G is elementary abelian of rank $\geqslant 3$, $\mathrm{Ch}(G)$ is properly contained in $H^{\text{even}}(G, \mathbb{Z})$.

The book falls into three parts: the cohomology of discrete groups (Chapters 1–4), representations, bundles and characteristic classes

(Chapters 5–6) and applications (Chapters 7–9). In the first part we base our treatment on that first given by J.P. Serre in *Corps locaux*, but emphasise those topics such as Frobenius reciprocity and the description of the image of the restriction homomorphism in terms of stable elements, which we exploit later. Chapter 4 is devoted to the spectral sequence of a group extension and to the use of this in calculating the integral cohomology in certain cases. The problem with an iterative application of the spectral sequence, say to p-groups, is that in general it does not collapse. However even in unfavourable circumstances the theory of characteristic classes may be used to identify universal cycles, thus simplifying some of the earlier calculations in the literature. For an algebraist this part of the book will seem relatively complete.

In Chapter 4 we summarise what we need from the theory of representations over an algebraically closed field, and from the theory of complex vector bundles. In neither case do we attempt to give more than a bare outline of proofs. In defining the Chern classes of a complex vector bundle we start from 1-dimensional or line bundles, and use the isomorphism between $\mathrm{Vect}^1_{\mathbb{C}}(X)$ and $H^2(X, \mathbb{Z})$ to define c_1. The algebraist may well prefer to start from the universal classes as dual to the homology classes carried by the Schubert varieties in the Grassmann manifold $G_{n,k}$ of n-dimensional subspaces in $\mathbb{C}^k$ (k large). In Chapter 6 we extend the theory of representations from $\mathbb{C}G$-modules over a point to families of $\mathbb{C}G$-modules indexed by the points x of a parameter space X on which G acts. Besides defining the kth Chern class of a representation ρ to be the kth, Chern class of the associated flat bundle over $BG = K(G, 1)$, we also use these classes to clarify some of the calculations in Chapter 4.

The third section on applications opens with a discussion of the symmetric group S_n. We prove that stably the subring generated by all Chern classes is actually generated by the classes $c_k(\pi_n)$ of the permutation representation. If $n \gg k$ and k is even, the order of $c_k(\pi_n)$ divides the denominator of B_k/k where B_k is the kth Bernoulli number. Here we adopt the convention that $B_k/k!$ equals the coefficient of t^k in the expansion of $e^t/(e^t - 1) + t/2$. The group S_n is extreme in the sense that, although its integral cohomology is finitely generated over $\mathrm{Ch}(S_n)$, the number of module generators increases rapidly with n, see [Mn]. One cause for this phenomenon is the existence of elementary abelian subgroups of large rank in S_n, and our second application avoids this problem by restricting the p-rank of a finite group to be at most 2.

If $rk_p(G) = 1$ the p-torsion in $H^*(G, \mathbb{Z})$ is periodic and generated by Chern classes; if $rk_p(G) = 2$ the best results are obtained for groups of prime power order (classified by N. Blackburn for $p \geqslant 5$). Our best theorem in this direction is that if G is a p-group which is either split metacyclic, or a central extension of $\mathbb{Z}/p^{n-2}$ by $(\mathbb{Z}/p) \times (\mathbb{Z}/p)$, then $H^{\text{even}}(G, \mathbb{Z}) = \text{Ch}(G)$. However it is clear that the method used extends to other classes of groups.

Finally we combine the theory for groups of low rank with the work of D. Quillen on linear groups over fields of finite characteristic containing sufficiently many roots of unity. This leads to a theory of characteristic classes for representations of G in the algebraically closed field $\bar{\mathbb{F}}_p$. By way of example note that our earlier methods show that

$$\text{Ch}(\text{SL}(3, \mathbb{F}_p))_{(l)} = H^{\text{even}}(\text{SL}(3, \mathbb{F}_p), \mathbb{Z})_{(l)}$$

if $l \geqslant 5$ and $l \nmid p(p-1)$. If l divides $p-1$ the l-torsion can be calculated by restriction to the subgroup D of diagonal matrices, Theorem 9.5, and if $l = p$ from the theory for groups of order p^3 (Theorem 8.6 and Appendix 2).

The book ends with a purely topological appendix, proving the Riemann–Roch theorem for group representations. This is motivated by the result in algebraic geometry – heuristically the multiple $M_k s_k$ of the Newton polynomial in the Chern classes $c_1 \ldots c_k$ arises from clearing fractions both from components of the Chern character and from the coefficients of the Todd genus. It is possible to give a purely algebraic proof of this result, with the sharper bound $\bar{M}_k =$ product of *distinct* primes dividing M_k, see [E–K2], but mathematically the more illuminating approach is that used here, which combines the modern treatment of natural maps between cohomology theories with transfer with elegant calculations by J.F. Adams.

With the exception of Appendix 1 prerequisites for reading this book are basic courses in algebraic topology, homological algebra and group theory. Chapter 5 will present no problems to the reader acquainted with J.P. Serre, *Représentations linéaires des groupes finis* [Se2] and D. Housemoller, *Fibre Bundles* [Hs].

I wish to thank several people: B. Eckmann, who has provided advice and inspiration over a number of years, J.F. Adams, M. Kervaire and B. Kahn, who have discussed the general theory with me and made valuable suggestions, M. Taylor, who as devil's advocate read Chapters 4–6, students and colleagues who attended a seminar on the applications in the final chapters held at Cambridge during the Michaelmas Term 1983. Among them I wish to thank J. Greenlees for the loan of his notes on the Riemann–Roch formula, also Gwen Jones and Rahel Boller for typing various parts of the manuscript.

I wrote the final version while a visitor at the Eidgenössische Technische Hochschule in Zürich (Summer 1983) and at the University of Geneva (Spring 1984); I am deeply grateful for the hospitality shown to me by both institutions.

Zürich and Cambridge

1

Group cohomology

Preliminaries

Let G be a discrete, not necessarily finite group. Denote by Λ the integral groupring $\mathbb{Z}G$ of G, consisting of formal sums $\sum n_i g_i (n_i \in \mathbb{Z}, g_i \in G)$ with the operations

$$\sum n_i g_i + \sum m_i g_i = \sum (n_i + m_i) g_i$$

and

$$(\sum n_i g_i)(\sum m_j g'_j) = \sum (n_i m_j)(g_i g'_j).$$

We shall be concerned with the abelian category $\mathfrak{A}_G$ of left Λ-modules and Λ-homomorphisms. The Λ-module A may be thought of as being defined by an abelian group A together with a homomorphism from G into $\mathrm{Aut}(A)$ – in short a G-action on A. We shall frequently refer to A as a G-module and write A_0 for the underlying $\mathbb{Z}$-module or abelian group. Denote by

$$A^G = \{a \in A : ga = a \text{ for all } g \in G\}$$

the subset of invariant elements. With these conventions we may define a G-action on $\mathrm{Hom}(A_0, A'_0)$ by the rule

$$(gf)(a) = gf(g^{-1}a),$$

from which it is clear that

$$\mathrm{Hom}_G(A, A') = \mathrm{Hom}(A_0, A'_0)^G.$$

Dually we define the diagonal G-action on $A \otimes A'$ by $g(a \otimes a') = ga \otimes ga'$. Note that both for homomorphisms and for tensor products with respect to the underlying commutative ring $\mathbb{Z}$ we omit the ring from the notation.

As in all categories of modules we have projective and injective objects; however these are too restrictive for our purposes, and their homologically significant properties are shared by coinduced and induced modules. Thus

Definition

Let X be an abelian group with trivial G-action (i.e. the image of G in $\mathrm{Aut}(A)$ is the identity). A is said to be coinduced if

$$A = \mathrm{Hom}(\Lambda, X),$$

and induced if

$$A = \Lambda \otimes X.$$

Exercise

If G is finite show that the notions of coinduced and induced modules coincide.

The technical usefulness of such modules is shown by

Lemma 1.1

Every G-module A embeds in a coinduced module.

Proof. Consider the map

$$A \to \mathrm{Hom}(\Lambda, A_0) \text{ given by } a \longmapsto f_a,$$

where $f_a(1) = a$ and f_a is extended to all of Λ by linearity. One checks easily that this map is (1–1) and is compatible with the G-actions.

Dually the tensor product $\Lambda \otimes A_0$ has the module A as homomorphic image – map $g \otimes a$ to ga. To see that this map is an epimorphism consider the splitting as abelian groups given by $a \longmapsto 1 \otimes a$.

Lemma 1.2

The map $A \to A^G$ is left exact. Thus if

$$A \underset{\phi}{\rightarrowtail} B \underset{\psi}{\twoheadrightarrow} C$$

is a short exact sequence of left Λ modules, the sequence

$$A^G \underset{\phi^G}{\rightarrowtail} B^G \underset{\psi^G}{\to} C^G$$

is exact in the category of abelian groups.

Proof. The only point which is not obvious is to show that $\operatorname{Ker}\psi^G \subseteq \operatorname{Image}\phi^G$. Let $b \in B^G$ with $\psi b = 0$. Then $b = \phi a$ for some $a \in A$. However

$$\phi(ga) = g(\phi a) = gb = b,$$

so ga and a have the same image under the monomorphism ϕ. It follows that a is an invariant, as required.

In general ψ^G is not an epimorphism, and this fact motivates the definition of the first cohomology group $H^1(G, A)$. Formally let us fix the group G and allow A to run through the objects of the category $\mathfrak{A}_G$.

Definition

The cohomology groups $H^k(G, A), k \geqslant 0$, form a covariant family of functors from $\mathfrak{A}_G$ to abelian groups, which has the following properties:

(1) $H^0(G, A) = A^G$,

(2) For each short exact sequence $A \rightarrowtail B \twoheadrightarrow C$ in $\mathfrak{A}_G$ there exists a natural transformation $\delta = \delta^k \colon H^k(G, C) \to H^{k+1}(G, A)$ and a long exact sequence of cohomology groups

$$\cdots \to H^k(G, A) \xrightarrow{\phi^*} H^k(G, B) \xrightarrow{\psi^*} H^k(G, C) \xrightarrow{\delta} H^{k+1}(G, A) \to \cdots.$$

(3) If A is a coinduced module, then $H^k(G, A) = 0$ for all $k \geqslant 1$.

In short the family $\{H^k(G, \cdot), k \geqslant 0\}$ is a cohomological extension of the invariant element functor, which vanishes on coinduced modules.

Theorem 1.3

The cohomological extension $\{H^k(G, \cdot), k \geqslant 0\}$ exists and is unique.

Proof. If $\mathbb{Z}$ is given the trivial G-structure, then $A^G = \operatorname{Hom}_G(\mathbb{Z}, A)$. This is so since any G-homomorphism is determined by the image of $1 \in \mathbb{Z}$. Now let

$$\cdots \to P_k \to P_{k-1} \to \cdots P_0 \twoheadrightarrow \mathbb{Z}$$

be an exact sequence of modules over the group ring Λ with each module P_k projective. Such a sequence is called a projective resolution of the trivial G-module $\mathbb{Z}$, and a specific example will be given below. By a standard argument in homological algebra any two such are chain homotopy equivalent. The composition of two successive homomorphisms in the related sequence

$$\cdots \leftarrow \operatorname{Hom}_G(P_k, A) \leftarrow \operatorname{Hom}_G(P_{k-1}, A) \leftarrow \cdots \leftarrow \operatorname{Hom}_G(P_0, A) \leftarrow A^G$$

is zero for any G-module A. Thus we may define $H^k(G,A) = H_k(\mathrm{Hom}_G(P_*, A))$. The existence of the natural coboundary homomorphisms δ^k and the long exact sequence of cohomology groups (2) follow by the usual diagram chase, and property (1) is satisfied, once we make the convention that the 'boundaries' in dimension 0 are trivial.

Consider the coinduced module $\mathrm{Hom}(\Lambda, X)$ for which

$$\mathrm{Hom}_G(P_k, \mathrm{Hom}(\Lambda, X)) \cong \mathrm{Hom}(P_{k,0}, X).$$

Since the G-projective module P_k is free over $\mathbb{Z}$, applying $\mathrm{Hom}(\cdot, X)$ to the projective resolution preserves exactness. Hence

$$H^k(G, \mathrm{Hom}(\Lambda, X)) = 0 \quad \text{for} \quad k \geqslant 1.$$

Uniqueness is obvious in dimension zero. Inductively assume that we have proved this up to dimension $k-1$, and consider the short exact sequence of coefficients in $\mathfrak{A}_G$

$$A \rightarrowtail \mathrm{Hom}(\Lambda, A_0) \twoheadrightarrow \bar{A}.$$

By properties (2) and (3)

$$\delta^{k-1}: H^{k-1}(G, \bar{A}) \cong H^k(G, A)$$

is an isomorphism, so that uniqueness also holds in dimension k. Note that in dimension 1 δ is only an epimorphism, but this is sufficient.

The technique just used to extend a result or construction from dimension 0 to dimension k is called *dimension shifting.* We shall use it frequently in what follows.

In order to complete the proof of (1.3) it remains to define the standard resolution for an arbitrary discrete group. Let $\bar{P}_k$ be the free $\mathbb{Z}$-module with basis given by $(k+1)$-tuples $(g_0, \ldots, g_k)$ of elements from G, and let G act via

$$g(g_0, \ldots, g_k) = (gg_0, \ldots, gg_k).$$

As basis elements with respect to the ring Λ rather than $\mathbb{Z}$ we may take $(k+1)$-tuples with $g_0 = 1$, the identity element of G. The boundary homomorphism $\mathrm{d}: \bar{P}_{k-1}$ is defined on each free generator over $\mathbb{Z}$ by

$$\mathrm{d}(g_0, \ldots, g_k) = \sum_{j=0}^{k} (-1)^j (g_0, \ldots \hat{g}_j \ldots g_k),$$

when as usual we read $\hat{g}_j$ as 'omit the element g_j'. In dimension zero we use the augmentation map $\varepsilon: P_0 \to \mathbb{Z}$ with $\varepsilon(g_0) = 1$. Formally we have borrowed the definition of d from that of the simplicial boundary of a simplex with vertices indexed by the g_j. Hence, since this simplex is acyclic, the algebraic sequence of Λ-modules and Λ-homomorphisms

$$\cdots \to \bar{P}_k \xrightarrow[d_k]{} \bar{P}_{k-1} \xrightarrow[d_{k-1}]{} \cdots \to \bar{P}_0 \xrightarrow[\varepsilon]{} \!\!\!\to \mathbb{Z}$$

is exact, and constitutes a free resolution of $\mathbb{Z}$ over Λ. Note in passing that when G is finite, this construction shows that each module $\bar{P}_k$ may be taken to be finitely generated.

A cochain in $\mathrm{Hom}_\Lambda(\bar{P}_k, A)$ may be identified with a function

$$f: G \times \cdots \times G \to A,$$

which satisfies the equivariance condition

$$f(gg_0, \ldots, gg_k) = gf(g_0, \ldots, g_k).$$

An equivariant cochain is thus determined by restriction to elements of the form $[g_1|\cdots|g_k] = (1, g_1, g_1g_2, \ldots, g_1g_2\ldots g_k)$, from which it follows that we may interpret the elements of $\mathrm{Hom}_\Lambda(\bar{P}_k, A)$ as non-homogeneous cochains f (on k arguments), for which the coboundary d^*f is given by the formula

$$\begin{aligned} d^*f(g_1, \ldots, g_{k+1}) = {} & g_1 f(g_2, \ldots, g_{k+1}) - f(g_1g_2, g_3, \ldots, g_{k+1}) \\ & + f(g_1, g_2g_3, g_4, \ldots, g_{k+1}) - \cdots \\ & + (-1)^j f(g_1, \ldots, g_jg_{j+1}, \ldots, g_{k+1}) + \cdots \\ & + (-1)^{k+1} f(g_1, \ldots, g_k). \end{aligned}$$

Exercise

Use this formula to check directly that $d^{*2} = 0$. Show also that it is possible to confine attention to the subcomplex of normalised cochains which satisfy the condition that $f(g_1, \ldots, g_k) = 0$ whenever some $g_j = 1$.

Although the standard resolution is important for the abstract definition of the groups $H^k(G, A)$, it is almost useless as a tool in calculations. These are best done by means of a special resolution for the group concerned, often motivated by topological considerations, and, as already noted, chain homotopy equivalent to the standard resolution above. For example let $G = C_r^T$, a cyclic group of order r generated by T, $2 \leqslant r \leqslant \infty$.

1. If $r = \infty$, one has the resolution

$$\mathbb{Z}C_\infty^S \rightarrowtail \mathbb{Z}C_\infty^T \twoheadrightarrow \mathbb{Z},$$

in which $S \to (T-1)$ and $T \to 1$.

Here the topological motivation is the elementary fact that the exponential map from the real numbers $\mathbb{R}$ to the circle S^1 is a universal covering map. The cyclic group C^T_∞ acts on $\mathbb{R}$ by mapping the half-open interval $[n, n+1)$ according to the rule $Tx = (x+1)$. For an arbitrary G-module A

$$H^0(G, A) = A^G,\ H^1(G, A) = A/(T-1)A,\ H^k(G, A) = 0,\ k \geqslant 2.$$

Thus C^T_∞ is an example of a group of cohomological dimension 1.

2. If $r < \infty$, write $N = 1 + T + T^2 + \cdots + T^{r-1}$, the sum of the group elements. Then by inspection the following is a free resolution of $\mathbb{Z}$ over $\mathbb{Z}C^T_r$ – topologically we consider an equivariant cellular decomposition of a sphere with respect to an induced linear action:

$$\cdots \rightarrow \Lambda_{(2k)} \xrightarrow[N]{} \Lambda_{(2k-1)} \xrightarrow[T-1]{} \cdots \xrightarrow[N]{} \Lambda_{(1)} \xrightarrow[T-1]{} \Lambda_{(0)} \longrightarrow \mathbb{Z}.$$

For an arbitrary G-module A it follows that

$$H^0(G, A) = A^G,\ H^{2k}(G, A) = A^G/NA,\ H^{2k-1}(G, A) = \operatorname{Ker} N/(T-1)A,\ k \geqslant 1.$$

The finite cyclic group C^T_r is an example of a group with *periodic cohomology*, a topic which we will systematically study in Chapter III below.

Low-dimensional interpretation

The formula for the coboundary shows that a 1-cocycle is a map $f: G \rightarrow A$ which satisfies the condition $f(g_1 g_2) = g_1 f(g_2) + f(g_1)$. Such a map is called a crossed homomorphism (note that when the G-action on A is trivial, a crossed homomorphism is a homomorphism in the usual sense). The coboundaries in dimension 1 are the principal crossed homomorphisms of the form $f_a(g) = ga - a$ for all $g \in G$. Thus $H^1(G, A)$ is isomorphic to the abelian group of crossed homomorphisms modulo principal crossed homomorphisms.

Similarly a 2-cocycle from the standard resolution is a map

$$f: G \times G \rightarrow A$$

which satisfies the condition

$$g_1 f(g_2, g_3) - f(g_1 g_2, g_3) + f(g_1, g_2 g_3) - f_1(g_1, g_2) = 0.$$

Such a function is called a factor system for the following reason. Consider the family of groups E which are extensions of the abelian group A_0 by G, and for which the G-structure on A corresponds to the action of G on A_0 by conjugation. This correspondence depends on the choice of a transversal $s: G \to E$ (set of coset representatives) for A_0, in G, and such a transversal satisfies

$$s(g_1)s(g_2) = f(g_1, g_2)s(g_1 g_2).$$

By messy but straightforward calculation one then shows

(i) the factor system f determines the composition law in E,

(ii) the cocycle identity above is equivalent to associativity in E, and

(iii) choice of a new transversal $s': G \to E$ changes f by a coboundary.

Hence the second cohomology group $H^2(G, A)$ describes the family of extensions

$$A_0 \rightarrowtail E \twoheadrightarrow G$$

for a specified G-action on A. Put another way the extension groups associated with the pair (G, A), where A is abelian as a normal subgroup, are determined up to isomorphism by the module structure on A and a 2-dimensional cohomology class. For a more leisurely discussion the reader is referred to the book [Mac].

Exercises

1. Under what conditions on A is $H^1(C^T_\infty, A) = 0$?

2. Using the calculation of $H^2(C^T_r, A)$ given above, determine all possible extensions of $\mathbb{Z}$ by $\mathbb{Z}/2$, of $\mathbb{Z}/p$ by $\mathbb{Z}/2$ (p = prime), and of $\mathbb{Z}/4$ by $\mathbb{Z}/2$. (As in the general discussion we adopt the convention that the first named group corresponds to the normal subgroup in E.)

As a further illustration of extension theory consider the familiar classification of groups of order p^3, where p is an odd prime number. Such a group is either an extension of $C^A_{p^2}$ by C^B_p or of $C^A_p \times C^C_p$ by C^B_p; in both cases we must determine first the possible module structures and then the size of H^2. When the normal subgroup is cyclic, and the module structure is trivial, the possible extensions are $C^A_{p^2} \times C^B_p$ (zero element in H^2) and $C^B_{p^3}$ (generator of $H^2(C_p, \mathbb{Z}/p^2) \cong \mathbb{Z}/p$). If the module structure is non-trivial – recall that Aut(C_{p^2}) is cyclic of order $p(p-1)$ – the extension is P_1, the unique non-abelian metacyclic group of order p^3. The group H^2 is trivial, since both the invariant elements and the image of N are isomorphic to $p\mathbb{Z}/p^2\mathbb{Z}$. When the normal subgroup is non-cyclic, and the module structure is trivial, we obtain $C^A_p \times C^B_p \times C^C_p$ or $C^A_{p^2} \times C^B_p$ (counted twice). Finally, if the module structure is trivial, which implies that the generator $\bar{B}$ is mapped to a parabolic element in $GL(2, \mathbb{F}_p)$, H^2 again vanishes and the extension defines the non-abelian group P_2 of order p^3 and exponent p.

The groups of order $p^4 (p \geqslant 5)$ are listed in Appendix 3. By considering such a group as an extension of a group of order p^3 by a cyclic group of order p, it is possible to prove that this list is exhaustive. However the simple extension theory described here must be generalised to allow for a non-abelian kernel, see [Gb] or [Br, IV.6]. One of the main additional ingredients needed is the identification of the elements of order p in the group of outer automorphisms of P_1 and P_2, which correspond to genuine extensions.

Homology groups

If A is a G-module let A_G be the quotient group of A by the subgroup generated by elements of the form $ga - a$. This quotient is sometimes called the group of coinvariants of A; it is the largest quotient group of A on which G acts trivially. In a similar way to Lemma 1.2 one may show that A_G is right exact, and by arguing as in Theorem 1.3 with the complex $\{P_k \otimes_\Lambda A : k \geqslant 0\}$ one obtains the homology groups $H_k(G, A)$. Note that, since P_k and A are both given a left structure, we must first define a right structure on P_k using the rule $xg = g^{-1}x$. The homology groups are unique and satisfy the following properties:

1. $H(G, A) = A_G$,
2. For each short exact sequence $A \rightarrowtail B \twoheadrightarrow C$ in $\mathfrak{A}_G$ there exists a natural transformation $\delta = \delta_k : H_k(G, C) \to H_{k-1}(G, A)$ and a long exact sequence of homology groups similar to that for cohomology.
3. If X is an abelian group, $H_k(G, \Lambda \otimes X) = 0$ for all $k \geqslant 1$, that is, the functors H_k are trivial on induced modules.

In principle the groups $H_k(G, A)$ may be calculated using the complex $\{\bar{P}_k \otimes_\Lambda A : k \geqslant 0\}$ associated to the standard resolution. An element $x \in \bar{P}_k \otimes_\Lambda A$ may be identified with the function $x(g_1, \ldots, g_k)$ taking values in A, which vanish almost everywhere. The boundary d_* is given by the formula

$$\begin{aligned} d_* x(g_1, \ldots, g_{k-1}) &= \sum_{g \in G} g^{-1} x(g, g_1, \ldots, g_{k-1}) \\ &+ \sum_{j=1}^{k-1} (-1)^j \sum_{g \in G} x(g_1, \ldots, g_j g, g^{-1}, g_{j+1}, \ldots, g_{k-1} \\ &+ (-1)^k \sum_{g \in G} x(g_1, \ldots, g_{k-1}, g). \end{aligned}$$

The most important low-dimensional interpretation is then given by

Lemma 1.4
Let $\mathbb{Z}$ have the trivial G-module structure and $[G, G]$ denote the commutator subgroup of G. Then $H_1(G, \mathbb{Z}) \cong G/[G, G]$.

Proof. As in the definition of the resolution for a cyclic group let $\varepsilon: \mathbb{Z}G \to \mathbb{Z}$ be the augmentation homomorphism with kernel I_G equal to the subgroup of $\mathbb{Z}G$ generated by the elements $i_g = g - 1$. With this notation

$$H_0(G, A) = A/I_G A.$$

From the short exact sequence which defines I_G we see that $H_0(G, I_G) = I_G/I_G^2$ with trivial image in $H_0(G, \Lambda)$. Since the group ring Λ is certainly projective, $H_1(G, \Lambda) = 0$ and we have an isomorphism

$$d_*: H_1(G, \mathbb{Z}) \xrightarrow{\simeq} H_0(G, I_G) = I_G/I_G^2.$$

The homomorphism $G \to I_G/I_G^2$ defined by $g \longmapsto i_g$ has kernel equal to $[G, G]$, from which the lemma follows.

Complete resolutions and the Tate groups

With the same notation as before multiplication by N, the sum of the group elements, defines an endomorphism $N: A \to A$ for any G-module A. Note that at this point we must restrict our attention to finite groups.

Clearly $I_G A \subseteq \operatorname{Ker} N$ and $\operatorname{Image} N \subseteq A^G$, and so N induces a homomorphism of abelian groups $N^*: H_0(G, A) \to H^0(G, A)$. Define ${}_N A$ to be the kernel of the operation of N on A, and

$$\hat{H}_0(G, A) = \operatorname{Ker} N^* = {}_N A/I_G A,$$

$$\hat{H}_0(G, A) = \operatorname{Coker} N^* = A^G/N A.$$

Lemma 1.5
If A is induced or coinduced, then $\hat{H}_0(G, A) = \hat{H}^0(G, A) = 0$.

Proof. Assume the result of the exercise following the definition of (co)induced modules, and restrict attention to $\hat{H}^0$. We may suppose that

$$A = \coprod_{g \in G} gX$$

for a suitable subgroup X of A_0. Since each $a \in X$ may be expressed uniquely as

$$a = \coprod g x_g,$$

a is an invariant if and only if all the x_g are equal, that is $a = Nx$ for some $x \in X$.

Therefore $A^G = NA$ and $\hat{H}^0(G, A) = 0$.

Lemma 1.6

If $A \rightarrowtail B \twoheadrightarrow C$ is a short exact sequence in $\mathfrak{A}_G$, *the diagram below is commutative with exact rows.*

$$\begin{array}{ccccccc}
H_1(G,C) \xrightarrow{d_*} & H_0(G,A) & \longrightarrow & (G,B) & \twoheadrightarrow & H_0(G,C) & \\
& \downarrow N_A^* & & \downarrow N_B^* & & \downarrow N_C^* & \\
& H^0(G,A) & \rightarrowtail & H^0(G,B) & \longrightarrow & H^0(G,C) & \xrightarrow[d^*]{} H^1(G,A)
\end{array}$$

Proof. This is immediate from the definitions.

A standard argument from homological algebra (3×3 Lemma) shows that there is a connecting homomorphism $\eta\colon \operatorname{Ker}(N_C^*) \to \operatorname{Coker}(N_A^*)$. The same argument or an easy diagram chase shows that η may be used to splice together homology and cohomology into a long exact sequence, extending to infinity in both directions:

$$\cdots \to \hat{H}_1(G,C) \xrightarrow[d^*]{} \hat{H}_0(G,A) \to \hat{H}_0(G,B) \to \hat{H}_0(G,C) \xrightarrow[\eta]{} \hat{H}^0(G,C)$$
$$\to \hat{H}^0(G,B) \to \hat{H}^0(G,A) \xrightarrow[d^*]{} H^1(G,C) \to \cdots$$

This justifies the definition of the Tate cohomology groups $\{\hat{H}^k(G,A)\colon -\infty < k < \infty\}$ as

$$\hat{H}^k(G,A) = H^k(G,A), \quad k \geqslant 1,$$
$$\hat{H}^0(G,A) = A^G/NA, \quad \hat{H}^{-1}(G,A) = N^A/I_G A,$$
$$\hat{H}^k(G,A) = H_{-k-1}(G,A), \quad k \leqslant -2.$$

It follows from this definition that for all G-modules A and for all $k \in \mathbb{Z}$ there are isomorphisms

$$\hat{H}^k(G,A) \cong \hat{H}^{k-1}(G, \operatorname{Hom}(\Lambda, A_0)/A) \text{ and } \hat{H}^k(G,A) \cong \hat{H}^{k+1}(G,K),$$

K is the kernel of the projection map $\Lambda \otimes A_0 \to A$ defined after Lemma 1.1.

Remark

The obvious period two in the cohomology groups $H^*(C_r^T, A)$, $r < \infty$, extends to the Tate groups.

Instead of defining the groups $\hat{H}^k$ in the rather *ad hoc* way adopted here, one can combine the complexes used to calculate homology and cohomology into a single *complete resolution* as follows: Splice together the projective resolution P_* and its abelian group dual $\mathrm{Hom}(P_{*0}, \mathbb{Z})$ obtaining

$$\begin{array}{ccccc} \cdots \to P_k \to \cdots \to P_0 & \longrightarrow & \mathrm{Hom}(P_{0,0}, \mathbb{Z}) \to \cdots \to \mathrm{Hom}(P_{k,0}, \mathbb{Z}) \to \cdots \\ & {\scriptstyle\varepsilon}\searrow \quad \nearrow{\scriptstyle\eta} & \\ & \mathbb{Z} & \end{array}$$

Apply $\mathrm{Hom}_\Lambda(\cdot, A)$, and use the isomorphisms $\psi: \mathrm{Hom}(B_0, \mathbb{Z}) \to \mathrm{Hom}_G(B, \Lambda)$ (given by $\psi(u)(b) = \sum_{g \in G} u(g^{-1}b)g$), $\psi: B_\Lambda \otimes A \to \mathrm{Hom}_\Lambda(\mathrm{Hom}_\Lambda(B, \Lambda), A)$ (given by $\psi(b \otimes a)(u) = au(b)$), taking care to observe the usual conventions about right and left structures and their interchange. The doubly infinite resolution becomes the doubly infinite complex

$$\begin{array}{ccccccc} \cdots \to P_0 \bigotimes_\Lambda A & \xrightarrow[\varepsilon_*]{} & \mathbb{Z} \bigotimes_\Lambda A & \dashrightarrow & \mathrm{Hom}_\Lambda(\mathbb{Z}, A) & \xrightarrow[\varepsilon^*]{} & \mathrm{Hom}_\Lambda(P_0, A) \to \cdots. \\ & & \| & & \| & & \\ & & A_G & \xrightarrow[N^*]{} & A^G & & \end{array}$$

The homology groups of this complex (with a shift of 1 in the labelling of negative dimensions) are then the Tate cohomology groups, for more detail see [C–E, Chapter XII, §3]. Note that one technical advantage of extending the definition of cohomology groups from the range $0 \leqslant k < \infty$ to $-\infty < k < \infty$ is that we can shift dimensions in either direction, starting from some property in dimension k_0.

Notes and references

In this introductory chapter I have more or less followed the treatment in [Se1]. For a more leisurely exposition of the basic concepts, particularly in low dimensions, see [Mac], especially Chapter IV. This book is also a useful reference for the necessary homological algebra. The definition of the Tate cohomology groups in the last section may be generalised from finite groups to infinite discrete groups of *virtually finite dimension.* These are groups defined by short exact sequences.

$$G_1 \rightarrowtail G \twoheadrightarrow Q,$$

where Q is finite and $H^k(G_1, A) = 0$ for all $k > n$ and all G_1-modules A. If there exists a module B such that $H^n(G_1, B) = 0$, n is called the (virtual) cohomological dimension of (G) G_1. By combining a complete resolution $\{Y_k: -\infty < k < \infty\}$ for the finite quotient group Q with a resolution $\{P_l: 0 \leqslant l < \infty\}$ of $\mathbb{Z}$ over G, which is such that

$$\mathring{P}_n = \operatorname{Ker}(P_n \to P_{n-1})$$

is projective *over* $\mathbb{Z}G_1$, it is possible to define groups $\hat{H}^k(G, A)$ for all integral values of k. The complete resolution for G is $\{Y_k \otimes \mathring{P}_n: -\infty < k < \infty\}$, each module of which is projective *over* $\mathbb{Z}G$. The groups $\hat{H}^k(G, A)$ can be studied by means of the lattice of finite subgroups of G, for example, they are annihilated by the lowest common multiple of the indices of the torsion free finite index subgroups G_1, and

$$\hat{H}^k(G_1, A) = H^k(G, A) \quad \text{for } k > n.$$

In particular a finite group is a group of virtual cohomological dimension 0. If in addition G_1 satisfies a form of duality then there is a long exact sequence connecting $\hat{H}^k$, H^k and H_k. This extended theory of complete resolutions is due to T. Farrell [F], and is developed at length in Chapter X of the book by K. Brown [Br]. Examples to which the theory applies include arithmetic groups such as $\mathrm{SL}(n, \mathcal{O})$ and $\mathrm{Sp}(2n, \mathcal{O})$, where $\mathcal{O}$ is the ring of integers in a finite extension field of the rational numbers $\mathbb{Q}$.

Problems

1. If G is a finite group (or more generally an infinite discrete group with torsion) show that $\mathbb{Z}$ does not admit a projective resolution of finite length over $\mathbb{Z}G$.
   [Hint: first consider the case when G is a non-trivial finite cyclic group.]
2. Let G be a finite group. Use the technique of dimension shifting to show that for each value of k there exists some G-module C such that $H^k(G, C)$ is cyclic of order $[G:1]$. Note that $\hat{H}^0(G, \mathbb{Z}) \cong \mathbb{Z}/[G:1]$.
3. Let G be the dihedral group of order 6, $G = \langle A, B: A^3 = B^2 = 1, A^B = A^2 \rangle$. Use the following diagram of free G-modules and homomorphisms to construct a resolution of $\mathbb{Z}$ over $\mathbb{Z}G$.

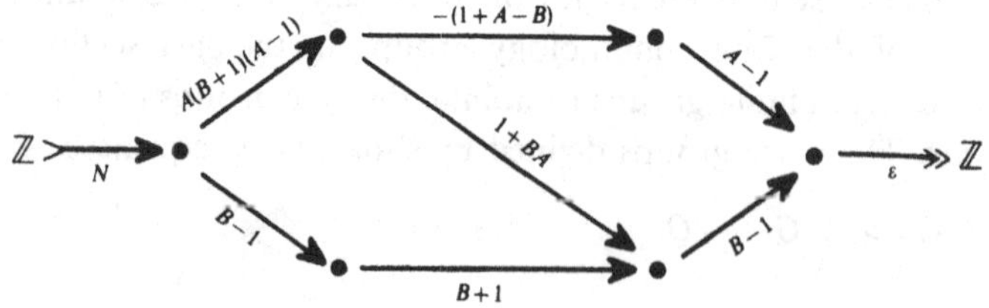

This resolution is evidently periodic with period 4, see the Appendix to R.G. Swan, Periodic resolutions for finite groups, *Annals of Math.* **72** (1960), 290.

4. Under what conditions on the coefficient module A does the group $H^1(C_\infty, A)$ vanish, that is, when is every crossed homomorphism principal?
5. [Harder] Read the section on extension theory with non-abelian kernel in [Br] or [Gb]. Calculate the outer automorphism group $\mathrm{Out}(P_1)$, where P_1 is the metacyclic group of order p^3, and fix some homomorphism $\psi: C_p \to \mathrm{Out}(P_1)$. Show that all extensions

$$1 \to P_1 \to E \to C_p \to 1$$

associated with ψ are determined by the elements of $H^2(C_p, \zeta(P_1))$, and compare your answer with the list in Appendix 3. Here $\zeta(P_1)$ denotes the centre of the group P_1.

2

Products and change of group

From now on we restrict attention to finite groups, although some of the results on change of group hold for subgroups of finite index in an infinite discrete group. Recall from the first chapter that $A \otimes B$ denotes the tensor product over the underlying ring $\mathbb{Z}$, and is given the diagonal G-structure, i.e. $g(a \otimes b) = ga \otimes gb$.

Definition of the cup product

Let Y_* be a complete resolution for G, for example the resolution obtained by splicing together (across the trivial module $\mathbb{Z}$ in dimension $-\frac{1}{2}$) the standard projective resolution and its dual over $\mathbb{Z}$. If we look at the invariant elements in the tensor product $A \otimes B$, it is clear that

$$A^G \otimes B^G \hookrightarrow (A \otimes B)^G,$$

and if $a \in A^G$, $b = Nb' \in NB$, then

$$a \otimes b = a \otimes Nb' = N(a \otimes b') \in N(A \otimes B).$$

Interchanging the roles of a and b we obtain a product map in dimension 0

$$\begin{array}{ccccc} \hat{H}^0(G, A) \otimes \hat{H}^0(G, B) & \to & \hat{H}^0(G, A \otimes B) \\ \| \qquad\qquad \| & & \| \\ A^G/NA \qquad B^G/NB & & (A \otimes B)^G/N(A \otimes B) \end{array}$$

The shift of this construction to other dimensions depends on the following rather technical result

Lemma 2.1

For each pair of integers (p, q) there exists a Λ-homomorphism $\phi_{p,q}: Y_{p+q} \to Y_p \otimes Y_q$, which satisfies

(i) $(\varepsilon \otimes \varepsilon)\phi_{0,0} = \varepsilon$, *and*

(ii) $\phi_{p,q}\mathrm{d} = \mathrm{d}'\phi_{p+1,q} + (-1)^p \mathrm{d}''\phi_{p,q+1}$

Here d′ and d″ are the differentials restricted to the first and second factors respectively of $Y_* \otimes Y_*$ with an appropriate convention as to signs.

Proof. Over $\mathbb{Z}$ the abelian group Y_k splits as $\operatorname{Ker} \mathrm{d}_k \oplus \operatorname{Im} \mathrm{d}_k$; so there exists a family of abelian group homomorphisms $s_k: Y_k \to Y_{k+1}$, such that

$$\mathrm{d}_{k+1}s_k + s_{k-1}\mathrm{d}_k = 1.$$

(The family s_k is a contracting homotopy.) Furthermore, since $Y_{k,0}$ is free, there exists an abelian group map $\rho: Y_k \to Y_k$ satisfying $N\rho = 1$. Following [C–E, X.8.6] for the construction of ρ we note that given the definition of the G-action on $\Lambda \otimes Y_k$, the natural projection onto Y_k splits via the G-map v. Since the elements $g \in G$ form a $\mathbb{Z}$-basis for Λ,

$$v(y) = \sum_{g \in G} g \otimes h(g, y).$$

The splitting map v is compatible with the group action, so

$$h(g, y) = gh(1, g^{-1}y),$$

and we may set $\rho(y) = h(1, y)$.

Let s', s'', ρ', ρ'' have the obvious meanings. The existence of the map $\phi_{0,0}$ follows from the fact that Y is Λ-projective and the possibility of completing the diagram

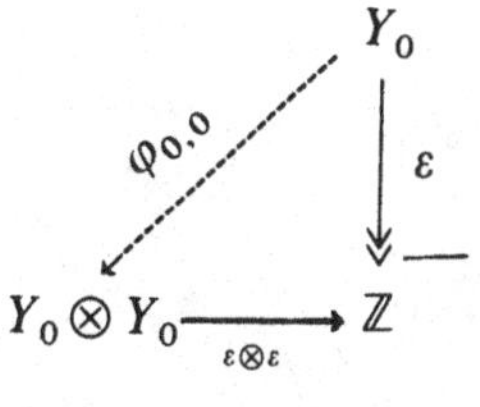

Now introduce the auxiliary condition

(iii) $\mathrm{d}'\phi_{p,q}\mathrm{d} = (-1)^p \mathrm{d}'\mathrm{d}''\phi_{p,q+1}$.

Step 1. Define homomorphisms $\phi_{0,q}$ which satisfy (i) and $(\mathrm{iii})_{(0,q)}$ for all q. Choose $z \in Y_0$ such that $\varepsilon(z) = 1$, and if $y \in Y_q$ set

$$\psi_{0,q}(y) = z \otimes y, \quad \text{and } \phi_{0,q} = N\psi_{0,q},$$

in order to have a G-invariant map. Note that

$$(\varepsilon \otimes 1)\phi_{0,q}(y) = 1 \otimes N\rho(y) = 1 \otimes y.$$

If $q = 0$,

$$\begin{aligned}(\varepsilon \otimes \varepsilon)\phi_{0,0}(y) &= (1 \otimes \varepsilon)(\varepsilon \otimes 1)\phi_{0,0}(y)\\ &= (1 \otimes \varepsilon)(1 \otimes y)\\ &= \varepsilon(y),\end{aligned}$$

so that the new definition of $\varphi_{0,0}$ agrees with the old. Now check $(\mathrm{iii})_{(0,q)}$:

$$\begin{aligned}\mathrm{d}'\mathrm{d}''\phi_{0,q+1}(y) &= \mathrm{d}''\mathrm{d}'\phi_{0,q+1}(y)\\ &= \mathrm{d}''(\mathrm{d} \otimes 1)\phi_{0,q+1}(y)\\ &= \mathrm{d}''(\mathrm{d} \otimes 1)N(z \otimes \rho y)\\ &= \mathrm{d}''(N\mathrm{d}z \otimes y)\\ &= N\mathrm{d}z \otimes \mathrm{d}y.\end{aligned}$$

Similarly

$$\begin{aligned}\mathrm{d}'\phi_{0,q}\mathrm{d}(y) &= \mathrm{d}'N(z \otimes \rho \mathrm{d}y)\\ &= N\mathrm{d}z \otimes \mathrm{d}y,\end{aligned}$$

agreeing with the first calculation.

Step 2. Define $\phi_{p,q}$ for $p < 0$. To be precise show that, if for some fixed negative integer p and *all* q there exist homomorphisms satisfying $(\mathrm{iii})_{p,q}$, then there exists a homomorphism $\phi_{p-1,q}$ satisfying $(\mathrm{ii})_{p-1,q}$. Since condition $(\mathrm{ii})_{p-1,q}$ is stronger than $(\mathrm{iii})_{p-1,q}$ this will provide the inductive step.

For fixed p and all q set $\phi_{p-1,q} = (-1)^p s''\mathrm{d}'\phi_{p,q-1}$. Then

$$\begin{aligned}\phi_{p-1,q}\mathrm{d} &= (-1)^p s''[(-1)^p \mathrm{d}'\mathrm{d}''\phi_{p,q}]\\ &= s''\mathrm{d}''\mathrm{d}'\phi_{p,q} = (1 - \mathrm{d}''s'')\mathrm{d}'\phi_{p,q}\\ &= \mathrm{d}'\phi_{p,q} - \mathrm{d}''((-1)^p\phi_{p-1,q+1})\\ &= \mathrm{d}'\phi_{p-q} + (-1)^{p-1}\mathrm{d}''\phi_{p-1,q+1},\end{aligned}$$

as required.

Step 3. Define $\phi_{p,q}$ for $p > 0$. Suppose that for some fixed positive integer p and all q there exist homomorphisms $\phi_{p,q}$ satisfying $(\mathrm{iii})_{p,q}$; then there exists a homomorphism $\phi_{p+1,q}$ satisfying $(\mathrm{ii})_{p,q}$. Since $(\mathrm{ii})_{p,q}$ and $(\mathrm{ii})_{p,q+1}$ together imply $(\mathrm{iii})_{p+1,q}$, the machinery to carry out an inductive construction is again available.

Now

$$\begin{aligned}(\mathrm{d}'s' + s'\mathrm{d}')\phi_{p,q}\mathrm{d} &= \mathrm{d}'s'\phi_{p,q}\mathrm{d} + (-1)^p s'\mathrm{d}'\mathrm{d}''\phi_{p,q+1}\\ &= \mathrm{d}'s'\phi_{p,q}\mathrm{d} + (-1)^p(1 - \mathrm{d}'s')\mathrm{d}''\phi_{p,q+1}\end{aligned}$$

$$= d'[s'\phi_{p,q}d + (-1)^p(s'd''\phi_{p,q+1})]$$
$$+ (-1)^p d''\phi_{p,q+1}.$$

In order to satisfy condition $(ii)_{p,q}$ we are therefore forced to set

$$\phi_{p+1,q} = s'\phi_{p,q}d + (-1)^{p+1}s'd''\phi_{p,q+1}$$

for all values of q. This completes the proof of the lemma.

At the level of cochains define the cup product by

$$f \cup h: Y_{p+q} \to A \otimes B, \quad \text{where } f \cup h = (f \otimes h)\phi_{p,q}.$$

Then $d^*(f \cup h) = d^*f \cup h + (-1)^p f \cup d^*h$, so that on passing to cohomology we obtain a bilinear map

$$\hat{H}^p(G, A) \otimes \hat{H}^q(G, B) \to \hat{H}^{p+q}(G, A \otimes B).$$

For the equivalence classes we write

$$[f \cup h] = [f]\cdot[h],$$

and this definition is natural in A and B. In dimension zero the cup product is induced by the natural map

$$A^G \otimes B^G \to (A \otimes B)^G.$$

Continuing the numbering from Lemma 2.1 and applying the techniques of homological algebra we can prove the following compatibility relations:

(iv) If $A_1 \rightarrowtail A_2 \twoheadrightarrow A_3$ and $A_1 \otimes B \rightarrowtail A_2 \otimes B \twoheadrightarrow A_3 \otimes B$ are short exact sequences in $\mathfrak{A}_G$, then the cup product is compatible with the coboundary homomorphism. The same is true (up to sign) if the roles of A and B are reversed.

Condition (iv) together with a dimension shifting argument shows that, subject to condition (ii) in Lemma 2.1, the cup product in group cohomology is uniquely defined. The same technique also shows that

(v) the product is associative (up to the identification of $(A \otimes B) \otimes C$ and $A \otimes (B \otimes C)$).

Furthermore

(vi) If $a \in \hat{H}^p(G, A)$ and $b \in \hat{H}^q(G, B)$, then $a\cdot b = (-1)^{pq} b\cdot a$ (up to the identification of $A \otimes B$ and $B \otimes A$).

This follows since $(-1)^{pq} b\cdot a$ is a product satisfying the same conditions as $a\cdot b$; now appeal to uniqueness.

A variant of the construction enables us to give $\hat{H}^*(G, A)$ the structure of a graded (anti)commutative ring, whenever A is a commutative ring with trivial G-action, for example $\mathbb{Z}$. If $\mu: A \times B \to C$ is G-invariant and $\mathbb{Z}$-bilinear, that is $\mu(ga, gb) = g\mu(a, b)$, then μ together with the cup product induces a bilinear map

$$\cdot_\mu: \hat{H}^p(G, A) \otimes \hat{H}^q(G, B) \to \hat{H}^{p+q}(G, C).$$

Exercise

[C–E, pages 251–2]. Let C_r^T be the cyclic group of order r, $r < \infty$, already considered in Chapter 1. The periodic projective resolution of $\mathbb{Z}$ constructed there can obviously be extended to $-\infty$, giving a complete resolution,

$$\cdots \to \Lambda \to \Lambda \to \Lambda \to \cdots.$$
$$T-1 \quad N \quad T-1 \quad N$$

We have already used this to determine the groups $\hat{H}^k(C_r^T, A)$. In order to compute products we need the maps

$$\phi_{p,q}: Y_{p+q} \to Y_p \otimes Y_q$$

for this particular resolution. These are given by

$$\phi_{p,q}(1) = \begin{cases} 1 \otimes 1, & p \text{ even} \\ 1 \otimes T, & p \text{ odd \& } q \text{ even} \\ \sum\limits_{0 \leqslant m < n \leqslant r-1} T^m \otimes T^n, & p \text{ odd \& } q \text{ odd.} \end{cases}$$

Hence, if $a \in \hat{H}^p(G, A)$ and $b \in \hat{H}^q(G, B)$, then as an element of $\hat{H}^{p+q}(G, A \otimes B)$

$$a \cdot b = \begin{cases} a \otimes b, & \text{one of } p \text{ or } q \text{ even,} \\ \sum\limits_{0 \leqslant m < n \leqslant r} T^m a \otimes T^n b, & p \text{ and } q \text{ both odd.} \end{cases}$$

In the important special case of $A = B = C = \mathbb{Z}$ with trivial G-action we shall simplify this result and give an easier proof as part of the discussion of periodicity in the next chapter.

Change of group

Let $\phi: G_1 \to G_2$ be a homomorphism of groups and A a G_2-module. The scalar product $g_1 a = \phi(g_1)a$ gives A the structure of a G_1-module, which we denote by $\varphi^! A$. If a is G_2-invariant, then a is also G_1-invariant, that is, ϕ induces a map between cohomology groups in dimension 0:

$$\phi^*: H^0(G_2, A) \to H^0(G_1, \phi^! A).$$

Either by direct construction of a chain map

$$\mathrm{Hom}_{G_2}(P_*^{(2)}, A) \to \mathrm{Hom}_{G_1}(P_*^{(1)}, \phi^! A)$$

or by appeal to a general argument in homological algebra ϕ^* extends to all positive dimensions

$$\phi^*: H^k(G_2, A) \to H^k(G_1, \phi^! A) \tag{1}$$

As a homomorphism between abelian groups ϕ^* is compatible with the coboundary homomorphism δ, and if there is no danger of confusion, for example when the module structure on A is trivial, we shall write $A = \phi^! A$.

More generally let us consider a G_1-module A_1 and a $\mathbb{Z}$-homomorphism $\alpha: A \to A_1$. The pair (ϕ, α) is *compatible* if

$$\alpha(\phi(g_1)a) = g_1\alpha(a) \quad \text{for all } g_1 \in G_1, a \in A.$$

Put another way (ϕ, α) is a compatible pair if $\alpha: \phi^! A \to A_1$ is a homomorphism of G_1-modules, which defines a family of maps

$$H^k(G_1, \phi^! A) \to H^k(G_1, A_1). \tag{2}$$

Composing the homomorphisms (1) and (2) in dimension k we have

$$(\phi, \alpha)^*: H^k(G_2, A) \to H^k(G_1, A_1) \tag{3}$$

Important examples of this construction are

1. Restriction. Let $i: K \to G$ be the inclusion of the subgroup K in G. Then by forgetting the action of elements of G outside K the G-module A becomes a K-module ($i^! A$). In this case the homomorphism (1) is called restriction,

$$i^*: H^k(G, A) \to H^k(K, A).$$

2. Inflation. Let K be a normal subgroup of G and write $\pi: G \to G/K$ for the projection homomorphism onto the quotient group. The invariant subgroup A^K is a (G/K)-module and the pair

$$(\pi: G \to G/K, A^K \hookrightarrow A)$$

is compatible, defining the inflation homomorphism

$$\pi^*: H^k(G/K, A^K) \to H^k(G, A).$$

There are similar constructions in homology – for reasons which will be clear when we discuss complete resolutions and induced homomorphisms for the Tate groups, the analogous map to (1) is called corestriction.

3. Conjugation. Let $G_1 = G_2$, $A_1 = A$ and $\phi: G \to G$ an inner automorphism $g \longmapsto hgh^{-1}$. Write $\alpha: A \to A$ for the $\mathbb{Z}$-homomorphism $a \longmapsto h^{-1}a$; this definition is chosen to ensure the compatibility of α and ϕ. The induced map between cohomology groups is denoted by c_h.

Lemma 2.2
The induced homomorphism c_h is the identity.

Proof. Obviously $c_h = 1$ in dimension zero; now shift dimensions.

Lemma 2.3
(Eckmann–Shapiro Lemma) *If $i: K \to G$ is an inclusion and A is a K-module, then $H^k(K, A) \cong H^k(G, \Lambda \otimes_K A)$.*

Proof. The module structure on $A \otimes A$ exists since Λ is a K-bimodule. In dimension zero one determines the invariant elements by writing $\Lambda \otimes_K A$ as the direct sum of copies of A, indexed and permuted by the elements of a transversal $1 = g_1, g_2, \ldots, g_{[G:K]}$. The invariant elements with respect to G are therefore contained in the invariant elements with respect to K in the first summand A_1. This may be identified with A^K. Again the general result follows by shifting dimensions.

Remark
In later chapters we shall have occasion to consider automorphisms induced on some normal subgroup K by inner automorphisms of the larger group G. As above we shall denote these by c_h, but it is clear that on the cohomology of K c_h need no longer be trivial.

4. Corestriction (in cohomology). This is a weak inverse to restriction, the definition of which as in the two previous lemmas starts in dimension 0. If $s = [G:K]$, let $1 = g_1, g_2, \ldots, g_s$ be a left transversal for K in G, and for each $a \in A^K$ set

$$N_{G/K}(a) = \sum_{i=1}^{s} g_i a.$$

Since a is K-invariant $N_{G/K}$ is independent of the transversal chosen, and the image belongs to A^G, that is, we have a homomorphism

$$N_{G/K} = i_*: H^0(K, A) \to H^0(G, A).$$

We extend the definition to dimension k by supposing this done up to dimension $k-1$, and then use the diagram

$$\begin{array}{ccc} H^k(K,A) & \xleftarrow{\delta} & H^{k-1}(K, \operatorname{Hom}(\Lambda, A_0)/A) \\ \Big\downarrow i_*^{(k)} & & \Big\downarrow i_*^{(k-1)} \\ H^k(G,A) & \xleftarrow{\delta} & H^{k-1}(G, \operatorname{Hom}(\Lambda, A_0)/A) \end{array}$$

to define i_* in dimension k. Note that Λ above is the group ring of the larger group G.

In homology there is a restriction homomorphism

$$H_k(G,A) \to H_k(K,A)$$

defined by using the homomorphism

$$N'_{G/K}(a) = \sum_{i=1}^{s} g_i^{-1} a$$

in dimension 0, and then extending as above. Note that, if g and g' belong to the same left coset gK, then the images of $g^{-1}a$ and $g'^{-1}a$ in A_K coincide. Furthermore the map $A \to A_K$ factors through A_G.

In dimension 1 this restriction map coincides with the classical 'transfer map' used in finite group theory. The dimension shift between 0 and 1 can be expressed in the diagram below – recall from Chapter 1 that $H_1(G, \mathbb{Z})$ is the abelianised group $G/[G,G]$:

$$\begin{array}{ccccc} G/[G,G] & \overset{\delta}{\rightarrowtail} & I_G/I_G^2 & & \\ & & & \overset{N'}{\searrow} & \\ \Big\downarrow \text{Res} & & & & I_G/I_K I_G \\ & & & \underset{\tau}{\nearrow} & \\ K/[K,K] & \underset{\delta}{\rightarrowtail} & I_K/I_K^2 & & \end{array}$$

N' is the map in dimension 0 just defined, τ is induced by the inclusion of $\mathbb{Z}K$ in $\mathbb{Z}G$, and the horizontal maps δ are both monomorphisms because the group rings are induced modules. Now let $g_1, \ldots, g_s$ be a *right* transversal for K in G; then

$$N'(g-1) = \sum_{i=1}^{s} g_i(g-1) \quad \text{modulo } I_K I_G.$$

For each i there exists some $j(i)$ such that $g_i g = x_i g_{j(i)}$ for some $x_i \in K$.

Hence, working modulo $I_K I_G$,

$$N'(g-1) = \sum_{i=1}^{s} x_i g_{j(i)} - \sum_{i=1}^{s} g_{j(i)}$$

$$= \sum_{i=1}^{s} (x_i - 1)g_{j(i)}$$

$$= \sum_{i=1}^{s} (x_i - 1).$$

This proves

Lemma 2.4
With the notation as above

$$\text{Res}: G/[G, G] \to K/[K, K]$$

is the abelianisation of the map $g \longmapsto \prod_{i=1}^{s} x_i$.

Exercise
[Se 1, page 130]. Let $\chi: K \to \mathbb{C}^*$ be a 1-dimensional representation of the subgroup K and let $i_!\chi$ be the induced s-dimensional representation of G, see Chapter 5 below. Then

$$\det(i_!\chi(g)) = \varepsilon(g)\chi(\text{Res}(g)),$$

where ε is the signature of the permutation of the cosets in G/K determined by g. Translated from homology to cohomology this equation gives the 1-dimensional integral Riemann–Roch formula, see Theorem 6.3 below and Appendix 1.

The restriction–corestriction terminology becomes clearer once one considers the Tate groups $\{\hat{H}^k(G, A): -\infty < k < \infty\}$. Since

$$N_G = N_{G/K} \circ N_K$$

$i^*: H^0(G, A) \to H^0(K, A)$ factors through the quotient group $\hat{H}^0$.

Lemma 2.5
i^* and i_* are homomorphisms of graded abelian groups, commuting with the coboundary homomorphism $\delta = \delta^k$. Restriction i^* (respectively corestriction i_*) is uniquely defined by its values in dimension 0

$$A^G/N_G A \to A^K/N_K A,$$

given by the inclusion of invariant elements (respectively

$$N_K^A/I_K A \to N_G^A/I_G A,$$

given by the projection of covariant elements).

Proof. Everything is clear except perhaps for the compatibility with $\delta^{-1}:\hat{H}^{-1}(\cdot, C) \to \hat{H}^0(\cdot, A)$. To this end consider the square

$$\begin{array}{ccc} {}_NC/I_GC & \xrightarrow{\delta} & A^G/NG \\ \downarrow i^* & & \downarrow i^* \\ {}_NC/I_KC & \xrightarrow{\delta} & A^K/NK \end{array}$$

Let $c \in {}_NC$ represent a class $\bar{c}$ in $\hat{H}^{-1}$. The class $\delta(\bar{c})$ is obtained by lifting c to b, applying N_G and then taking the class modulo the submodule generated by N_G. Hence, if we apply $i^*\delta$ to $\bar{c}$ we obtain the class of N_Gb modulo N_K. Going the other way (via δi^*) we must lift $\sum_{i=1}^s g_i c$ to $\sum_{i=1}^s g_i b$ and apply N_K, giving N_Gb (modulo N_K) again. There is a similar argument for corestriction.

Lemma 2.6

Let A be a commutative ring of coefficients. Then

(i) $i^*:\hat{H}^k(G, A) \to \hat{H}^k(K, A)$ *is a ring homomorphism, and*

(ii) $i_*:\hat{H}^k(K, A) \to \hat{H}^k(G, A)$ *is a homomorphism of $\hat{H}^*(G, A)$ modules.*

Proof. In the light of Lemma 2.5 it is enough to consider dimension zero, and then shift dimensions. In dimension zero both assertions are immediate. Note that in terms of elements (ii) may be written as

$$i_*(i^*(x)y) = x \cdot i_*(y),$$

the Frobenius reciprocity formula.

The following assertions also hold because they hold in dimension zero, and we may apply Lemma 2.5 again.

(1) If $[G:K] = s$, $i_* \cdot i^*$ equals multiplication by s.

(2) $[G:1]\hat{H}^k(G, A) = 0$ for all A and all $k \in \mathbb{Z}$.

(3) If A is a G-module of finite type over $\mathbb{Z}$, then $\hat{H}^k(G, A)$ is a finite group of exponent dividing the order of G.

Note concerning notation. Where there is no chance of ambiguity in succeeding chapters we shall use the symbol i^* for restriction. Where it is important to distinguish both the subgroup and the inclusion monomorphism we shall use

$$i^*_{G \to K} \text{ or } \mathrm{Res}_{G \to K}.$$

A similar convention applies to i_* and $\mathrm{Cor}_{K \to G}$.

Notes and references

The material contained in this chapter is standard. The use of the complete resolution to define products in all dimensions, positive, negative and zero is based on [C-E, Chapter XII]. Topologically this treatment includes both cap and cup products. The section on change of group homomorphisms is again based on the relevant sections in Serre's book *Corps locaux* [Se 1], where the reader will also find some discussion of corestriction for a subgroup of finite index in an infinite group.

Problems

1. Let M and N be left $\mathbb{Z}K$ and $\mathbb{Z}G$-modules, where K is a subgroup of the finite group G, and write $\bar{N}$ for N regarded as a $\mathbb{Z}K$-module. Give an explicit proof that

$$N \otimes (\mathbb{Z}G \bigotimes_K M) \cong \mathbb{Z}G \bigotimes_K (\bar{N} \otimes M),$$

where G acts diagonally on the left hand side and K acts diagonally on the bracketed term $\bar{N} \otimes M$.

2. Consider the extension

$$1 \to K \to E \to G \to 1$$

where K is abelian and the orders $[K:1]$ and $[G:1]$ are coprime. Show that $H^2(G, K) = 0$, where G acts on K by conjugation, and deduce that the extended group E is uniquely determined by this module structure. What can one say if K is non-abelian?

3. If $f: K \hookrightarrow G$ is a monomorphism of infinite discrete groups, it is clear how to define the restriction map $f^*: H^k(G, A) \to H^k(K, A)$. If K has finite index in G, show that is also possible to define the corestriction map $f_*: H^k(K, A) \to H^k(G, A)$, and compare its properties with those which hold in the finite case. Now consider the following special cases:
   (a) If A is a G-module such that $H^k(K, A) = 0$, then $[G:K]x = 0$ for all $x \in H^k(G, A)$.
   (b) If $G = \mathrm{SL}(2, \mathbb{Z})$, the special linear group of 2×2 matrices with integral entries, then $12x = 0$ for all $x \in H^k(G, \mathbb{Z})$, $k \geqslant 2$. (It is not hard to see that $\mathrm{SL}(2, \mathbb{Z})$ contains a free subgroup of index 12, and that the cohomological dimension of any free group equals one. Generalise the argument for C_∞ given in the text.)
   (c) Consider $\mathrm{SL}(3, \mathbb{Z})$ in the same way, and find positive integers l and m such that $lx = 0$ for all $x \in H^k(G, \mathbb{Z})$, $k \geqslant m$. (Hint: consider a suitable principal congruence subgroup, or see C. Soulé, The cohomology of $\mathrm{SL}(3, \mathbb{Z})$, *Topology* **17** (1978), 1–22, for more precise calculations.)

3

Relations with subgroups and duality

Relations between subgroups

Let K and K' be subgroups of the finite group G, and decompose G as the union of pairwise disjoint double cosets

$$G = \bigcup_i K g_i K'.$$

Write the image of the subgroup K under conjugation by the element g_i as $K^{g_i} = g_i K g_i^{-1}$, and write L_i for the intersection $K \cap K'^{g_i}$.

Lemma 3.1

(i) $[G:K'] = \sum_i [K:L_i]$,
(ii) $i^*_{G\to K} i_{*K'\to G} = \sum_i i^*_{L_i \to K} i_{*K'^{g_i}\to L_i} c_{g_i}$, *and*
(ii) *if* $K \lhd G$ *then* $i^*_{G\to K} i_{*K\to G}(x) = \sum_i c_{g_i} x$.

Proof. (i) Write the subgroup K as a union of disjoint left L_i cosets,

$$K = \bigcup_j h_{ji} L_i, \text{ so that}$$

$$K g_i = \bigcup_j h_{ji}(K g_i \cap g_i K'), \text{ and therefore}$$

$$K g_i K' = \bigcup_j h_{ji}(K g_i K' \cap g_i K')$$

$$= \bigcup_j h_{ji} g_i K'.$$

Taking the union over i exhausts the elements of G, that is $G = \bigcup_{i,j} h_{ji} g_i K'$, as required. In (ii) we show that the left and right sides of the equation coincide when the dimension equals zero, and then shift dimensions. With the same notation for coset representatives as in (i) the right hand side for invariant elements takes the form

$$A^{K'} \longrightarrow A^{K'^{g_i}} \longrightarrow A^{L_i} \longrightarrow A^K$$

$$a \longmapsto g_i a \longmapsto g_i a \longmapsto \sum_j h_{ji} g_i a,$$

followed by summation over i. Since the elements $h_{ji}g_i$ for varying i and j form a K'-transversal in G, and the final sum is interpreted as an element in A^K, the definition of $i_{*K'\to G}$ in dimension 0 shows that

$$i^*_{G\to K} i_{*K'\to G}(a) = \sum_{i,j} h_{ji} g_i a,$$

as claimed. Part (iii) follows immediately from part (ii), by taking $K = K' = K'^{g_i}$. Formula (ii) is often called the double coset formula.

Definition
The class $x \in \hat{H}^k(K, A)$ is said to be *stable* if

$$i^*_{K\to K^g\cap K}(x) = i^*_{K^g\to K\cap K^g} c_g(x) = c_g i^*_{K\to K\cap K^g}(x)$$

for all $g \in G$.

Note that if K is a normal subgroup of G, then the stable elements are precisely those invariant under the action of G/K.

Lemma 3.2
(i) *If x belongs to the image of $i^*_{G\to K}$, then x is stable.*
(ii) *If x is stable, then $i^*_{G\to K} i_{*K\to G}(x) = [G:K]x$*

Proof. (i) If $x = i^* y$, then

$$i^*_{K^g\to K\cap K^g}(c_g x) = i^*_{G\to K\cap K^g}(c_g y)$$
$$= i^*_{G\to K\cap K^g}(y),$$

because conjugation is trivial on the cohomology of G.

$$= i^*_{K\to K\cap K^g}(x),$$

by decomposing $i^*_{G\to K\cap K^g}$ as $i^*_{K\to K\cap K^g} i^*_{G\to K}$

(ii) Apply formula (ii) from Lemma 3.1 with $K = K'$. Then

$$i^*_{G\to K} i_{*K\to G}(x) = \sum_i i_{*L_i\to K} i^*_{K^{g_i}\to L_i}(c_{g_i} x)$$
$$= \sum_i i_{*L_i\to K} i^*_{K\to L}(x),$$

using the stability of x,

$$= \sum_i [K:L_i]x$$
$$= [G:K]x, \text{ by Lemma 3.1 (i).}$$

In the next theorem, so as to confine attention to finitely generated abelian groups, let the G-module A be of finite type over $\mathbb{Z}$. Write $\hat{H}^k(G,A)_{(p)}$ for the p-torsion subgroup of the finite abelian group $\hat{H}^k(G,A)$; this has order dividing the highest power of p dividing the order of G. Clearly we have a decomposition as a finite direct product of rings:

$$\hat{H}^*(G,\mathbb{Z}) \cong \prod_p \hat{H}^*(G,\mathbb{Z})_{(p)}.$$

Let $\{G_p\}$ be a representative family of p-Sylow subgroups of G, as p runs through the primes dividing the order of G.

Theorem 3.3
The following sequence of abelian groups is exact:

$$0 \to \hat{H}^k(G,A)_{(p)} \xrightarrow[\Phi]{} \hat{H}^k(G_p,A) \xrightarrow[\Psi]{} \prod_{\substack{G_p \cap G_p^g \\ g \in G}} \hat{H}^k(G_p \cap G_p^g, A),$$

where

$$\Phi(x) = i^*_{G \to G_p}(x), \Psi_g(x) = i^*_{G_p \to G_p \cap G_p^g}(x) - i^*_{G_p^g \to G_p^g \cap G_p}(c_g x)$$

Proof. In words rather than symbols the theorem asserts that the p-torsion summand of $\hat{H}^k(G,A)$ restricts isomorphically onto the stable subgroup of $\hat{H}^k(G_p,A)$.

Notation: write $[G_p:1] = p^\nu$, $[G:G_p] = s$, so that $(p^\nu, s) = 1$, and let l be chosen so that $sl \equiv 1$ modulo p^ν. The previous lemma shows that the image of the restriction map consists of stable elements. Conversely let y be stable, then

$$li^*_{G \to G_p} i_{*G_p \to G}(y) = sl(y) \text{ by Lemma 3.2 (ii),}$$
$$= y, \text{ by choice of } l.$$

Hence y lies in the image of $i^*_{G \to G_p}$.

Furthermore $li_{*G_p \to G} i^*_{G \to G_p}(x)$

$$= ls(x)$$
$$= x, \text{ if } x \text{ is chosen to lie in the } p\text{-torsion subgroup.}$$

Hence the restriction homomorphism is (1–1), and for the record corestriction is an epimorphism. The map Ψ has been defined in such a way to make the sequence exact.

Note first that if $G_p \lhd G$, then by a previous lemma there is an identification of $i^*\hat{H}^k(G, A)_{(p)}$ and $\hat{H}^k(G_p, A)^{G/G_p}$. Secondly

$$\hat{H}^*(G_p, A) \cong \mathrm{Ker}(i_{*G_p \to G}) \oplus \mathrm{Image}(i^*_{G \to G_p}).$$

Particularly in the discussion of groups with p-periodic cohomology, see below and Chapter 8, the next result gives a useful description of the stable elements.

Lemma 3.4 (R. G. Swan).

If the Sylow subgroup G_p of G is abelian, and N_p denotes its normaliser in G, then

$$\hat{H}^*(G, A)_{(p)} \cong \hat{H}^*(G_p, A)^{N_p}.$$

Proof. Let Q_p denote the group of automorphisms of G_p induced by inner automorphisms of G, that is, by conjugation by elements of N_p. We have to show that the element $x \in \hat{H}^k(G_p, A)$ is stable if and only if x is fixed under the action of Q_p. One way round this is immediate; for the other assume that x is fixed by Q_p. If we denote the inclusion of $G_p \cap G_p^g$ in G_p by i and the inclusion of $G_p \cap G_p^g$ in G_p^g, followed by conjugation by g as j_g, then we must show that

$$i^*x = j_g^*x.$$

If Z is the centraliser of $G_p \cap G_p^g$ in G, both and G_p and G_p^g are contained in Z, because G_p is abelian. Since G_p is *a fortiori* a maximal p-subgroup of Z, there exists an element h belonging to Z such that

$$hgG_pg^{-1}h^{-1} = G_p.$$

Since h centralises all elements f of $G_p \cap G_p^g$,

$$j_g(f) = j_g(h^{-1}fh) = g^{-1}h^{-1}fhg = j_{hg}(f).$$

Hence $j_g = j_{hg} \circ i$, and j_{hg} is defined by an element of the subgroup of automorphisms Q_p. By assumption on $x, j_{hg}^* = 1$ and the lemma is proved.

Duality

The main result of this section, Lemma 3.7, will be used to construct generators for periodic cohomology. Our discussion is an outline only, and if the reader is prepared to accept the lemma on trust – and it is used nowhere else in the book – then he should proceed directly to the next section. For complete details see [C–E, Chapter XII, §6].

Recall that for abelian groups there is the adjoint pairing

$$\alpha: \mathrm{Hom}(A\otimes B, C) \cong \mathrm{Hom}(A, \mathrm{Hom}(B, C))$$

and the evaluation homomorphism

$$\varepsilon: \mathrm{Hom}(B, C) \times B \to C,$$

the second of which is compatible with G-structure. Hence following the discussion of products in the previous chapter one obtains a unique map

$$\hat{H}^p(G, \mathrm{Hom}(B, C)) \otimes \hat{H}^q(G, B) \to \hat{H}^{p+q}(G, C),$$

denoted by

$$(x, y) \longmapsto xy.$$

After composition with α one obtains

$$h_{p,q}: \hat{H}^p(G, \mathrm{Hom}(B, C)) \to \mathrm{Hom}(\hat{H}^q(G, B), \hat{H}^{p+q}(G, C)),$$

given explicitly by

$$h_{p,q}(x)(y) = xy.$$

Lemma 3.5
Let the short exact sequence $B_1 \underset{i}{\rightarrowtail} B_2 \underset{j}{\twoheadrightarrow} B_3$ *be such that*

$$\mathrm{Hom}(B_3, C) \underset{j^*}{\rightarrowtail} \mathrm{Hom}(B_2, C) \underset{i^*}{\twoheadrightarrow} \mathrm{Hom}(B_1, C)$$

is also short exact. If $x_1 \in \hat{H}^p(G, \mathrm{Hom}(B_1, C))$, $y_3 \in \hat{H}^{q-1}(G, B_3)$, *then* $\delta x_1 \in \hat{H}^{p+1}(G, \mathrm{Hom}(B_3, C))$ *and* $\delta y_3 \in \hat{H}^q(G, B_1)$ *are such that*

$$(\delta x_1)y_3 + (-1)^p x_1 \delta y_3 = 0.$$

Proof. Represent the classes y_3 and x_1 by cocycles and go through the usual construction of a connecting homomorphism. This involves a choice of intermediate cochains, which by abuse of notation one may label as $y_2: Y_q \to B_2$ and $x_2: Y_p \to \mathrm{Hom}(B_2, C)$. (Here as in the previous chapter Y_* denotes some complete resolution for G.) Since the product of two cocycles is a cocycle

$$0 = (\mathrm{d}^* x_2)y_2 + (-1)^p x_2(\mathrm{d}^* y_2).$$

The first component $(d^*x_2)y_2 = (j^*x_3)y_2 = x_3(jy_2)$, which represents $(\delta x_1)y_3$. There is a similar argument for the second component. Lemma 3.5 implies that the diagram below commutes up to sign:

$$\begin{array}{ccc} \hat{H}^p(G, \mathrm{Hom}(B_1, C)) & \xrightarrow{h_{p,q}} & \mathrm{Hom}(\hat{H}^q(G, B_1), \hat{H}^{p+q}(G, C)) \\ \delta^* \downarrow & & \downarrow \mathrm{Hom}(\delta, 1) \\ \hat{H}^{p+1}(G, \mathrm{Hom}(B_3, C)) & \xrightarrow{h_{p+1,q-1}} & \mathrm{Hom}(\hat{H}^{q-1}(G, B_3), \hat{H}^{p+q}(G, C)). \end{array}$$

Lemma 3.6

For a fixed module C and varying module B the map $h_{p,q}$ is an isomorphism if and only if the map $h_{i,j}$ is an isomorphism for all $i + j = p + q$.

Proof. It is enough to consider the case $i = p + 1, j = q - 1$. The module $\Lambda \otimes B$ is induced and $\mathrm{Hom}(\Lambda \otimes B, C)$, being the sum of coinduced modules, is also coinduced. Apply the diagram above to the exact sequences of coefficients

$$I_G \otimes B \rightarrowtail \Lambda \otimes B \twoheadrightarrow B \quad \text{and} \quad \mathrm{Hom}(B, C) \rightarrowtail \mathrm{Hom}(\Lambda \otimes B, C) \twoheadrightarrow \mathrm{Hom}(I_G \otimes B, C)$$

Since the symbol " $\wedge$ " is reserved for Tate cohomology groups, denote the dual of the abelian group by $DB = \mathrm{Hom}(B, \mathbb{Q}/\mathbb{Z})$.

Lemma 3.7

The map

$$h_{0,-1}: \hat{H}^0(G, DB) \longrightarrow \mathrm{Hom}(\hat{H}^{-1}(G, B), \hat{H}^{-1}(G, \mathbb{Q}/\mathbb{Z}))$$

is an isomorphism.

Proof. Using the fact that $\mathbb{Q}/\mathbb{Z}$ is a divisible abelian group write $h_{0,-1}$ as

$$\begin{array}{ccc} \hat{H}^0(G, DB) & \xrightarrow{h_{0,-1}} & \mathrm{Hom}(\hat{H}^{-1}(G, B), \mathbb{Q}/\mathbb{Z}) \\ \| & & \| \\ \dfrac{\mathrm{Hom}_G(B, \mathbb{Q}/\mathbb{Z})}{N\,\mathrm{Hom}(B, \mathbb{Q}/\mathbb{Z})} & \dashrightarrow & \mathrm{Hom}(N(B/I_G B), \mathbb{Q}/\mathbb{Z}). \end{array}$$

If one chases back through the definition of $h_{0,-1}$, one sees that it is obtained by restricting a G-homomorphism from B to $\mathbb{Q}/\mathbb{Z}$ to the subgroup

${}_N B$. Given $f: {}_N B \to \mathbb{Q}/\mathbb{Z}$ vanishing on $I_G B$, divisibility implies that there exists an extension $f': B \to \mathbb{Q}/\mathbb{Z}$, also vanishing on $I_G B$ and hence belonging to $\mathrm{Hom}_G(B, \mathbb{Q}/\mathbb{Z})$. Therefore $h_{0,-1}$ is onto. If $f'' \in \mathrm{Hom}_G(B, \mathbb{Q}/\mathbb{Z})$ is such that $f''({}_N B) = 0$, then the exactness of the coefficient sequence at the left

$$ {}_N B \rightarrowtail B \xrightarrow[N]{} B $$

and the divisibility (i.e. injectivity) of $\mathbb{Q}/\mathbb{Z}$ together imply that on applying $\mathrm{Hom}(\cdot, \mathbb{Q}/\mathbb{Z})$ one obtains a sequence of coefficients, which is exact on the right:

$$ \mathrm{Hom}(B, \mathbb{Q}/\mathbb{Z}) \xrightarrow[N^*]{} \mathrm{Hom}(B, \mathbb{Q}/\mathbb{Z}) \twoheadrightarrow \mathrm{Hom}({}_N B, \mathbb{Q}/\mathbb{Z}). $$

Therefore there exists a homomorphism f''' such that $f'' = N f'''$, and

$$ (N f''')(b) = \sum_{g \in G} g f'''(g^{-1} b) = \sum_{g \in G} f'''(g^{-1} b). $$

Summing over g^{-1} rather than over g the last term gives $f'''(Nb) = f''(b)$. It follows that f'' actually belongs to $N\,\mathrm{Hom}(B, \mathbb{Q}/\mathbb{Z})$, so that $h_{0,-1}$ is also (1–1).

Since this argument is valid for an arbitrary G-module B, on applying Lemma 3.6 and shifting dimensions one deduces that

$$ h_{-p,p-1}: \hat{H}^{-p}(G, \mathrm{D}B) \xrightarrow{\simeq} D(\hat{H}^{p-1}(G, B)) $$

is an isomorphism. Suppose that B equals $\mathbb{Z}$ with the trivial G-action, so that $h_{-p,p-1}$ is an isomorphism between $\hat{H}^{-p}(G, \mathbb{Q}/\mathbb{Z})$ and $\mathrm{D}\hat{H}^{p-1}(G, \mathbb{Z})$. After composition with the connecting isomorphism associated with the sequence of coefficients $\mathbb{Z} \rightarrowtail \mathbb{Q} \twoheadrightarrow \mathbb{Q}/\mathbb{Z}$ one concludes that

$$ \hat{H}^{-p+1}(G, \mathbb{Z}) \cong \mathrm{D}\hat{H}^{p-1}(G, \mathbb{Z}) \cong \mathrm{Hom}(\hat{H}^{p-1}(G, \mathbb{Z}), \mathbb{Z}/[G:1]). $$

Explicitly the composition in the previous line says that for each homomorphism

$$ \varphi: H^{p-1}(G, \mathbb{Z}) \to \mathbb{Z}/[G:1] $$

there is a unique element $x \in \hat{H}^{-(p-1)}(G, \mathbb{Z})$ such that $\varphi(y) = x \cdot y$.

Periodicity

If A is an abelian group, we let $A_{(p)}$ denote the p-torsion subgroup.

Definition

The element $z_p \in \hat{H}^{k_0}(G, \mathbb{Z})_{(p)}$ is called a *p-generator* if it generates $\hat{H}^{k_0}$ as an abelian group and has order equal to the maximal power of p dividing $[G:1]$. When there is no chance of confusion drop the suffix p.

Theorem 3.8

The following properties are equivalent:

(i) *z is a p-generator,*

(ii) *z has order equal to the maximal power of p dividing* $[G:1]$,

(iii) *there exists $z^{-1} \in \hat{H}^{-k_0}(G, \mathbb{Z})_{(p)}$ such that $z \cdot z^{-1} = 1$, and*

(iv) *$\hat{H}^k(G, A)_{(p)} \cong \hat{H}^{k+k_0}(G, A)_{(p)}$ for all $k \in \mathbb{Z}$ and all G-modules A.*

Proof. (i) implies (ii) from the definition.

(iii) follows from (ii) by using the final isomorphism of the previous section with $k_0 = p - 1$, and restricting to p-torsion. Note that the inverse element z^{-1} is unique. Given (iii) the isomorphism (iv) is obtained by taking the cup product with z. Finally (iv) implies (i), since

$$\hat{H}^0(G, \mathbb{Z})_{(p)} \cong \hat{H}^{k_0}(G, \mathbb{Z})_{(p)} \cong \mathbb{Z}/[G_p:1],$$

where as in the previous section G_p is a representative p-Sylow subgroup of G.

If z_1 and z_2 are p-generators in dimensions k_1 and k_2, then $z_1 z_2$ is a p-generator in dimension $k_1 + k_2$. Hence all p-periods are multiples of some minimal p-period equal to d_p. Furthermore d_p is even and is called the *cohomological p-period* for the finite group G.

Lemma 3.9

(i) *If G has a p-period, then so does every subgroup K of G, and $i^*_{G \to K}(z_p)$ is a p-generator.*

(ii) *Let G_p be a p-Sylow subgroup of G and $z_p \in \hat{H}^k(G_p, \mathbb{Z})$ a* generator of order $p^\nu = [G_p:1]$. *Suppose that the integer r is chosen to satisfy the congruence $q^r \equiv 1 (p^\nu)$ for all q coprime with p. Then $z_p^r \in \hat{H}^{kr}(G_p, \mathbb{Z})$ is stable, and $i_{*G_p \to G}(z_p^r)$ has order p^ν. Therefore G has p-period dividing kr.*

Proof. (i) If $p^{\nu(G)}, p^{\nu(K)}$ denote the orders of representative p-Sylow subgroups of K and G respectively, consider the composition

$$i_{*K \to G} i^*_{G \to K} = \text{multiplication by } [G:K].$$

On p-torsion this is equivalent to multiplication by $p^{\nu(G)-\nu(K)}$, so $i^*_{G \to K}(z_p)$

has order bounded below by $p^{v(K)}$, hence order equal to $p^{v(K)}$.

(ii) If z_p is a p-generator for the subgroup G_p, then $c_g z_p$ is a p-generator for G_p^g. Hence by (i) $i^* z_p = z_1$ and $i^* c_g z_p = z_2$ are both p-generators for $G_p^g \cap G_p$. There exists an integer q such that $qz_1 = z_2$, so that $z_2^r = q^r z_1^r = z_1^r$, so that z_p^r is stable. By lemma 3.2(ii)

$$i^* i_* z_p^r = [G:G_p] z_p^r,$$

and the right hand side has order p^v since the index of G_p in G is not divisible by p. Hence $i_* z_p^r$ also has order p^v, and serves as a p-generator for G.

Theorem 3.10
Let p be an odd prime. The following conditions are equivalent:
(i) *G has a p-period* $d_p > 0$,
(ii) *every abelian p-subgroup of G is cyclic, and*
(iii) *every p-(Sylow) subgroup of G is cyclic.*

Proof. (i) implies (ii). The Künneth formula holds for the cohomology of finite groups, and

$$H^*(C_p \times C_p, \mathbb{F}_p) \cong H^*(C_p, \mathbb{F}_p) \otimes H^*(C_p, \mathbb{F}_p),$$

which contains an $\mathbb{F}_p$-polynomial subalgebra on two generators. Hence $C_p \times C_p$ has no p-period.

(ii) implies (iii). Otherwise a p-subgroup P (non-cyclic, hence non-abelian) would have to contain a non-abelian group of order p^3 (see Chapter I, page 10). However $C_p \times C_p$ is a subgroup of both P_1 and P_2, leading to a contradiction. (iii) implies (i). In Chapter I we showed that a cyclic group has a p-period for each prime dividing the order; now apply Lemma 3.9 (ii). Note that this part of the argument also shows that if G_2 is cyclic, then G has 2-periodic cohomology.

When the prime p is odd, it is easy to calculate the p-period. Let N_p be the normaliser of G_p in G and Z_p its centraliser.

Lemma 3.11
With the notation already established, if p is odd,

$$d_p = 2[N_p:Z_p].$$

Proof. By lemma 3.4 it suffices to find a generator of $\hat{H}^{d_r}(G_p, \mathbb{Z})$ invariant under the induced action of N_p.

$$H^2(G_p, \mathbb{Z}) \cong \mathrm{Hom}(G_p, \mathbb{Q}/\mathbb{Z}).$$

Let z_p be a p-generator for G_p and ζ_p the corresponding homomorphism, mapping z_p to $1/p^\nu$ say. Then

$$\zeta_p^g(z_p) = \zeta_p(c_g z_p) = \zeta_p(z_p^r) = r/p^\nu$$

for some value of r, not divisible by p. The smallest value of r such that $\zeta_p^g = \zeta_p$ equals $[N_p : Z_p]$, the order of the subgroup of automorphisms of G_p induced by the action of N_p/Z_p.

Note that this argument also applies to the case when G_2 is cyclic. The index $[N_2 : Z_2]$ is odd, because $Z_2 \supseteq G_2$, and $\mathrm{Aut}(G_2)$ has order equal to a power of two. Hence the image of N_2/Z_2 is trivial, from which it follows that the 2-period equals two.

Example. Let D_{pq} be the non-abelian group of order pq, where p and q are distinct primes and q divides $p-1$. Such a group has a presentation

$$\{A, B : A^p = B^q = 1, A^B = A^r, r^q \equiv 1 \pmod p\}.$$

Lemma 3.11 shows that the q-period equals 2 – note that D_{pq} retracts onto the subgroup generated by B. The p-period equals $2q$, and from this it is easy to write down the whole integral cohomology ring.

In order to complete the discussion of periodicity it is necessary to introduce the binary dihedral or generalised quaternion groups, D_{4t}^*, $t \geqslant 2$. The former name expresses the fact that D_{4t}^* is a central extension of C_2 by D_{2t}, the dihedral group, the latter that D_{4t}^* is isomorphic to the subgroup of the unit quaternions generated by $e^{\pi i/t}$ and j. In terms of generators and relations

$$D_{4t}^* = \{A, B : A^{2t} = 1, A^t = B^2, A^B = A^{-1}\}.$$

If t is odd, $\hat{H}^*(D_{4t}^*, \mathbb{Z})$ can be calculated using Lemma 3.11. In general the easiest way to proceed is by means of a free linear action by the group on the topological 3-sphere S^3. We now sketch the algebraic consequences of this as an extended example-exercise.

1. The diagram of boundary maps below defines part of a complete resolution for D_{4t}^* lying between dimensions $4k$ and $4k+4$. The modules are free of rank 1 in dimensions congruent to 0 and 3 (mod 4) and of rank 2 for 1 and 2 (mod 4). It is clear that this complete resolution is periodic,

and induces a periodicity isomorphism in cohomology for all p dividing $2t$, see [C–E Chapter XII].

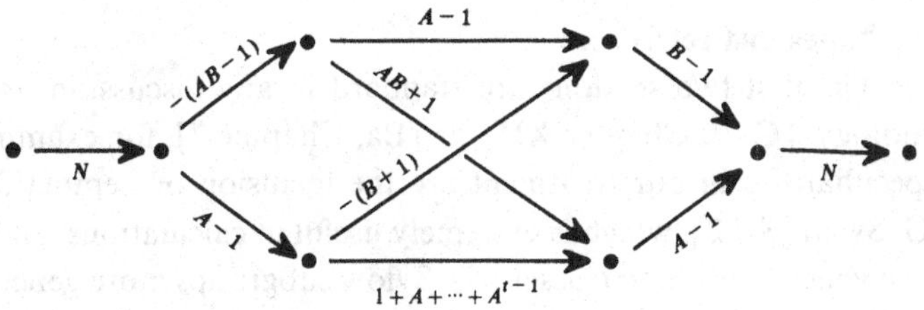

2. With coefficients in the trivial module $\mathbb{Z}$,

$\hat{H}^0 \cong \mathbb{Z}/4t$ (compare Theorem 3.8),

$\hat{H}^1 = H^3 = 0$, and

$$\hat{H}^2 \cong \begin{cases} \mathbb{Z}/4, & t \text{ odd} \\ (\mathbb{Z}/2) \times (\mathbb{Z}/2), & t \text{ even.} \end{cases}$$

As a ring $H^*(D^*_{2^n}, \mathbb{Z}) = \{\alpha, \beta, \xi: 2\alpha = 2\beta = 2^n\xi = 0,\ \alpha\beta = 0,\ \alpha^2 = \beta^2 = 2^{n-1}\xi\}$, $n \geqslant 4$. If $n = 3$, $\beta^2 = 0$ also.

3. If G_2 is a 2-group containing a unique subgroup of order 2, then either G_2 is cyclic or $G_2 \cong D^*_{2^n}$.

4. G has a 2-period if and only if either G_2 is cyclic or G_2 is binary dihedral.

5. Calculate the 2-period of G when $G_2 \cong D^*_{2^n}$, see Lemma 8.4 below and [Sw 2]. Note that Lemma 3.9 (ii) shows that an upper bound for the 2-period is given by the smallest positive integer $4r$ such that $q^r \equiv 1 \pmod{2^\nu}$ for all odd numbers q. Since $\text{Aut}(\mathbb{Z}/2^n) \cong \mathbb{Z}/2 + \mathbb{Z}/2^{n-2}$,

$$4r = 4 \cdot 2^{n-2} = 2^n$$

will do. However the more delicate calculation referred to shows that the 2-period actually equals four.

The group G is said to be (globally) periodic if and only if d_p exists for all primes dividing $[G:1]$, and in this case

$$d = \text{l.c.m.}(d_p), \quad \text{for} \quad p \mid [G:1]$$

The period d exists if and only if G_p is cyclic (p = odd) and G_2 is cyclic or binary dihedral. Topologically these groups arise in the study of free actions on spheres; algebraically they were first classified by Zassenhaus in the solvable case and by Suzuki otherwise, see the discussion and the bibliography in the book by J. Wolf [Wo]. The most important non-solvable examples are the special linear groups SL$(2, p)$, p an odd prime, See Chapter 9 below.

Notes and references

The first two sections are standard in any discussion on group cohomology, [C–E, Chapter XII] or [Ba, Chapter 7], for example. The only peculiarities in our treatment are the inclusion of Lemma 3.4, due to R.G. Swan [Sw 2], which is extremely useful in calculations, and which admits a generalisation to classes of p-Sylow subgroups more general than abelian. We state the conclusion of Theorem 3.3 as an exact sequence in order to emphasise its formal similarity with R. Brauer's theorem on induced representations, Theorem 5.4 below. The results on globally periodic groups are due to E. Artin and J. Tate; we have given the local form of their results, since many of the finite groups in which we are most interested possess a period d_p only for certain primes dividing the order.

Problems

1. Let G be a p-normal finite group, i.e. the centre of the p-Sylow subgroup G_p is the centre of each p-Sylow subgroup in which it is contained. Let N be the normaliser of the centre of G_p. Show that for any G-module A an element $x \in \hat{H}^k(G_p, A)$ is stable with respect to G if and only if it is stable with respect to N. (See the Appendix to R.G. Swan, the p-period of a finite group, *Ill. J. Math* **4** (1960), 345.)
2. Check that the diagram of free modules and homomorphisms given in the text for the binary dihedral group D^*_{4t} actually does define a free resolution of $\mathbb{Z}$ over $\mathbb{Z}D^*_{4t}$ of period 4.
3. Prove that G has cohomological period 2 if and only if G is cyclic. [Hint: consider $H_1\,(G, \mathbb{Z}) \cong \hat{H}^{-2}(G, \mathbb{Z})$.]
4. If G is a finite cyclic group and $A \rightarrowtail B \twoheadrightarrow C$ is a short exact sequence of coefficient modules, the long exact sequence of Chapter 1 reduces to an exact hexagon

$$\begin{array}{ccccc} & \nearrow & \hat{H}^0(G,A) \to \hat{H}^0(G,B) & \searrow & \\ \hat{H}^1(G,C) & & & & \hat{H}^0(G,C) \\ & \nwarrow & \hat{H}^1(G,B) \leftarrow \hat{H}^1(G,A) & \swarrow & \end{array}$$

Suppose that the groups with coefficients A are finite of order $h_i(A)$, $i = 0$, 1, and write $h(A) = h_0(A)/h_1(A)$. Establish the following properties of the rational number $h(A)$, known as the Herbrand quotient:
(a) $h(B) = h(A) \cdot h(C)$

(b) If A is a finite G-module, $h(A) = 1$.

(Consider the exact sequence $A^G \rightarrowtail A \xrightarrow{T-1} A \twoheadrightarrow A_G$, where T generates G, and then use the definitions of $\hat{H}^i$ $(i = 0, -1)$ given in Chapter 1.)

(c) If $f \in \mathrm{Hom}_\Lambda(A, B)$ has finite kernel and cokernel, then $h(A) = h(B)$.

(d) Extend the definition of the Herbrand quotient to the groups D_{pq} (of cohomological period $2(q-1)$) and prove the analogues of properties (a), (b) and (c).

4

Spectral sequences

The concern of the three previous chapters has been general theory – the only calculations have been for cyclic groups and for the closely related groups with periodic cohomology. The example of the non-abelian groups D_{pq} illustrates the roles played by a normal subgroup and by a quotient group on which a splitting map is defined. The simplicity of this example arises from the fact that the spectral sequence associated with a defining extension for D_{pq} is trivial away from the fibre and base. However even where this is no longer the case, and we are particularly interested in a not necessarily split extension of one p-group by another, careful examination of the E_2-page of the spectral sequence can sometimes give the structure of the cohomology of the extension group, at least as a graded abelian group.

The spectral sequence of an extension

Recall that a first quadrant spectral sequence in cohomology is a family $\{E_r^{i,j}: i \geqslant 0, j \geqslant 0\}$ of bigraded modules, the suffix r increasing from 1 or 2 to infinity, together with differentials

$$\mathrm{d}_r: E_r^{i,j} \to E_r^{i+r,j-r+1}$$

of bidegree $(r, 1-r)$, and isomorphisms $H(E_r^{i,j}, \mathrm{d}_r) \cong E_{r+1}^{i,j}$.
In the limit one defines $E_\infty^{i,j} = Z_\infty^{i,j}/B_\infty^{i,j}$, where

$$Z_\infty^{i,j} = \bigcap_r Z_r^{i,j} \quad \text{and} \quad B_\infty^{i,j} = \bigcup_r B_r^{i,j}$$

are the universal cycles (elements killed for all the differentials d_r) and coboundaries (elements hit by some differential d_r) respectively. Of particular interest are the terms with one superscript equal to 0. Thus, when

$i=0$ (fibre terms)

$$E_\infty^{0,j} = \bigcap_r \operatorname{Ker} d_r,$$

and when $j=0$ (base terms)

$$E_\infty^{i,0} = \bigcup_r E_r^{i,0}/\operatorname{Image} d_r.$$

One obtains such a spectral sequence from any differential graded abelian group with a compatible decreasing filtration F^i by defining

$$E_1^{i,j} = H^{i+j}(F^iA^*/F^{i+1}A^*),$$

with $B^k = \coprod_l B^{l,k-l}$, that is one introduces a single grading by taking the sum of the two degrees. In order to guarantee convergence, that is the existence of an isomorphism between the formally defined limit term $E_\infty^{i,j}$ and the quotient group $F^i(H^{i+j}A)/F^{i+1}(H^{i+j}A)$, assume that the filtration $\{F^0A = A, F^1A, \ldots\}$ satisfies the uniform finiteness condition

$$F^{k+1}A^k = 0 \text{ for all values of } k.$$

In rather more detail

$$E_r^{i,j} = (Z_r^{i,j} + F^{i+1}A^{i+j})/(dZ_{r-1}^{i-r+1,j+r-2} + F^{i+1}A^{i+j}),$$

laborious but straightforward calculation showing first that the condition for a spectral sequence is satisfied, and second that the finiteness assumption ensures convergence. The edge maps on the fibre and base – monomorphic for the former, epimorphic for the latter, see the definitions above – are induced by the natural maps

$$H^j(A) \to H^j(A/F^1A) \quad \text{and} \quad H^i(Z_1^{i,0}) \to F^iH^i(A) \hookrightarrow H^i(A).$$

Furthermore, if $i \geqslant 2$ there is a homomorphism defined on a submodule of $E_1^{0,i-1}$ taking values in a quotient module of $E_2^{i,0}$ called the transgression τ.

Let K be a normal subgroup of the discrete group G and A an arbitrary G-module. In a similar way to that in which we defined a G-module structure on $\operatorname{Hom}(A, B)$, see Chapter 1, we may define a (G/K)-module structure on the cohomology group $H^k(K, A)$. At the level of cochains let f belong to $\operatorname{Hom}_K(P_K, A)$ and write

$$(\bar{g}f)(x) = \bar{g}f(\bar{g}^{-1}x),$$

for each equivalence class $\bar{g}$ in G/K. Use the standard resolutions for G and

G/K to form the bigraded complex

$$B^{i,j} = \mathrm{Hom}_{G/K}(Y_i(G/K), \mathrm{Hom}_K(Y_j(G), A)).$$

By the properties of adjoint functors together with the definition of the (G/K)-structure on $\mathrm{Hom}_K(Y_j(G), A)$ just given the right hand side is isomorphic to

$$\mathrm{Hom}_G(Y_i(G/K) \otimes Y_j(G), A).$$

Filter this double complex in each of the two obvious ways:

$$(F_1^i B)_k = \coprod_{h \leqslant i} B^{h,k-h} \quad \text{and} \quad (F_2^i B)_k = \coprod_{h \leqslant i} B^{k-h,h},$$

with associated boundary homomorphisms $\mathrm{d}^{(1)}$ and $\mathrm{d}^{(2)}$. Each filtration is associated with a spectral sequence which converges to

$$H^*(\mathrm{Tot}\,(B)),$$

where the differential $\mathrm{d} = \mathrm{d}^{(1)} + \mathrm{d}^{(2)}$ on the total complex maps from total degree k into total degree $k+1$. For the first spectral sequence

$$E_{(1),2}^{i,j} \cong H_1^i H_2^j(B^{**});$$

for the second reverse the order of differentiation. This follows immediately from the definitions.

Lemma 4.1
$H^k(G/K, \mathrm{Hom}_K(\mathbb{Z}G, A)) = 0, \ k > 0.$

Proof. This generalises the statement for coinduced modules (take $K = \{1\}$) used in Chapter 1. The modification of the argument is left as an exercise.

Consider $(H_1^j(B))_i = H^j(G/K, \mathrm{Hom}_K(Y_i(G), A))$. Since the module $Y_i(G)$ is free the coefficients $\mathrm{Hom}_K(Y_i(G), A)$ have the vanishing property of Lemma 4.1, that is

$$(H_1^j(B))_i = 0 \quad \text{for} \quad j > 0.$$

Furthermore $(H_1^0(B))_i = \mathrm{Hom}_K(Y_i(G), A)^{G/K} = \mathrm{Hom}_G(Y_i(G), A)$.
Now taking homology with respect to $\mathrm{d}^{(2)}$:

$$H_2^i(\mathrm{Hom}_G(Y_i(G), A)) = H^i(G, A) = E_{(2),2}^{i,0},$$

and $E_{(2),2}^{i,j} = 0$ for $j > 0$. Since the only non-zero terms lie on the base the differentials must be trivial, and $E^{i,j}_{(2),2} = E^{i,j}_{(2),\infty}$. Equivalently

$$H^k(G, A) \cong H^k(\mathrm{Tot}(B)),$$

the isomorphism being induced by a chain map

$$\phi\colon \mathrm{Hom}_G(Y_*(G), A) \to \mathrm{Hom}_G(Y_0(G/K) \otimes Y_*(G), A),$$

given by $\phi f(\bar{g} \otimes x) = f(x)$. Turning to the first spectral sequence with $E^{i,j}_{(1),2} = H^i_1 H^j_2(B)$, we may calculate the cohomology of the subgroup K with coefficients in A (restricted) by using the standard resolution $Y_*(G)$, the modules of which are *a fortiori* K-projective.

Now

$$B^{i,*} = \mathrm{Hom}_{G/K}(Y_i(G/K), \mathrm{Hom}_K(Y_i(G), A))$$

splits as a direct sum corresponding to the splitting of $Y_i(G/K)$ as a free (G/K)-module. Therefore

$$H^j_2(B^{i,*}) = \mathrm{Hom}_{G/K}(Y_i(G/K), H^j(K, A)),$$

where $H^j(K, A)$ is to be regarded as a (G/K)-module. Apply the first boundary map $\mathrm{d}^{(1)}$, and obtain the cohomology of G/K. Since both spectral sequences converge to the cohomology of the total complex, this proves:

Theorem 4.2
Given a short exact sequence of groups

$$K \rightarrowtail G \twoheadrightarrow G/K$$

and an arbitrary G-module A, there is a first quadrant spectral sequence $\{E^{i,j}_r, \mathrm{d}_r\colon 2 \leqslant r < \infty, 0 \leqslant i, j < \infty\}$, which is natural in A, such that

$$E^{i,j}_2 \cong H^i(G/K, H^j(K, A)),$$

and which converges to a graded group associated to $H^*(G, A)$.

On the fibre and base there are isomorphisms

$$E^{i,0}_2 \cong H^i(G/K, A^K) \quad \text{and} \quad E^{0,j}_2 \cong H^j(K, A)^{G/K} = H^j(K, A)^G.$$

The edge maps may be identified with restriction and inflation respectively, and as a homomorphism the transgression τ has domain a subgroup of $H^{i-1}(K, A)$ and image a quotient group of $H^i(G/K, A^K)$ for all $i > 1$.

Exercise

Verify the description of the edge maps just given. This is not quite routine, see [Mac, page 353].

Remark.

Up to sign the isomorphism $E_2^{i,j} \cong H^i(\cdot, H^j(\cdot))$ is compatible with the bigraded ring structure.

Corollary 4.3

(i) *For an arbitrary short exact sequence of groups* $K \rightarrowtail G \twoheadrightarrow G/K$, *the induced sequence of cohomology groups*

$$0 \to H^1(G/K, A^K) \xrightarrow[\text{Inf}]{} H^1(G, A) \xrightarrow[\text{Res}]{} H^1(K, A)^G \xrightarrow[\tau]{}$$
$$H^2(G/K, A^K) \xrightarrow[\text{Inf}]{} H^2(G, A)$$

is also exact.

(ii) *With the notation of* (i) *suppose that for some* $m \geqslant 1$

$$H^j(K, A) = 0 \quad \textit{for} \quad 0 < j < m.$$

Then

$$\text{Inf}\colon H^j(G/K, A^K) \to H^j(G, A)$$

is an isomorphism for $j < m$,

$$\tau\colon H^m(K, A) \to H^{m+1}(G/K, A^K)$$

has domain as large as possible, and there is an exact sequence as in (i) *with* m *replacing* 1.

Proof. Part (i) follows from a description of the terms of total degree 1 and of some of the terms of total degree 2 in $E_2^{i,j}$, which survive to infinity. Explicitly one has the exact sequence

$$0 \to E_2^{1,0} \to H^1(G, A) \to E_2^{0,1} \xrightarrow[\tau]{} \underbrace{E_2^{2,0} \twoheadrightarrow E_3^{2,0} \to H^2(G, A)}_{\text{inf}},$$

in which $E_2^{1,0} = E_\infty^{1,0}$, $\operatorname{Ker}\tau = E_\infty^{0,1}$, and $E_3^{2,0} = E_\infty^{2,0}$, because of the increasing length of the differentials.

For part (ii) the entries on the $E_2^{i,j}$ page are trivial in the horizontal strip determined by $1 \leqslant j \leqslant m-1$. Inflation is an isomorphism for $j < m$, since $H^j(G/K, A)$ is the only non-zero entry on the line of total degree j in this range. In degree m there is an additional contribution coming from the image of the restriction map in $H^m(K, A)^G = E_2^{0,m}$, and in degree $m+1$ inflation maps $E_2^{m+1,0}/\text{Image}(\tau)$ into $H^{m+1}(G, A)$.

Part (i) may be proved directly, see [Se 1, Chapter VII, §6], by chasing representative cocycles. Part (ii) follows by shifting dimensions from m to 1.

An application – split metacyclic groups

The following result is a trivial consequence of Theorem 4.2:

Theorem 4.4

If $K \rightarrowtail G \twoheadrightarrow G/K$ is a (necessarily split) short exact sequence of finite groups such that $([K:1], [G:K]) = 1$, then for each $k > 0$ and each coefficient module A, there is a split short exact sequence of abelian groups:

$$H^k(G/K, A^K) \underset{\text{Inf}}{\rightarrowtail} H^k(G, A) \underset{\text{Res}}{\overset{\longleftarrow}{\twoheadrightarrow}} H^k(K, A)^G.$$

Proof. From the numerical formulae following Lemma 2.6 an element in $E_2^{i,j}$ $(i > 0, j > 0)$ has order dividing both the order and the index of K, hence must equal zero. Furthermore the only possible non-zero differential is

$$\tau: E_k^{0,k-1} \to E_k^{k,0},$$

but since the domain and image have coprime orders, τ vanishes also. Therefore $E_2^{i,j} = E_\infty^{i,j}$ and there can be only two non-zero terms on each line of total degree k. These have coprime orders, so the extension problem is trivial, giving the result.

Theorem 4.4 combined with the calculation of the cohomology groups of the cyclic group C_r in Chapter 1 determines the cohomology of G, when this is the semi-direct product of two cyclic groups of coprime orders. Examples are provided by the groups D_{4t}^* $(t = \text{odd})$ and D_{pq} discussed at the end of the last chapter – the reader should compare the result obtained there using the stable subgroup in $H^*(G_p, \mathbb{Z})$ with the prediction of the spectral sequence argument above.

Somewhat more complicated than the example just mentioned is the case of a split metacyclic p-group. For simplicity let us assume that p is odd, that G contains a cyclic normal subgroup of index p, and that the coefficient module is $\mathbb{Z}$. The method is however quite general – see [Wa]. The group G has a presentation

$$\{A, B: A^{p^a} = B^p = 1, A^B = A^r, r = p^{a-1} + 1 \,(\text{mod}\, p^a)\},$$

where we suppose that $a \geqslant 2$ in order to have a non-abelian group. In what follows $\langle A \rangle$ denotes the cyclic subgroup generated by the element A.

Consider the spectral sequence of the extension with

$$E_2^{i,j} = H^i(\langle B \rangle, H^j(\langle A \rangle, \mathbb{Z})).$$

Since $\langle B\rangle$ is cyclic, taking the cup product with an element $\beta \in H^2(\langle B\rangle, \mathbb{Z})$ of order p defines isomorphisms

$$\hat{E}_2^{0,j} \cong E_2^{2i,j}, \quad E_2^{1,j} \cong E_2^{2i+1,j}.$$

In order to study the invariant elements $H^j(\langle A\rangle, \mathbb{Z})^B = H^0(\langle B\rangle, H^j(\langle A\rangle, \mathbb{Z}))$ represent $\alpha \in H^2(\langle A\rangle, \mathbb{Z})$ by means of the character

$$\alpha\colon \langle A\rangle \to \mathbb{Q}/\mathbb{Z}, \quad \text{which is such that}$$
$$\alpha(A) = 1/p^a.$$

Then $\alpha^B(A) = \alpha(A^B) = r/p^a$, so that $\alpha^B = r\alpha$.
More generally and abbreviating the cup product of α with itself i times as α^i, we have

$$(\alpha^i)^B = r^i\alpha^i = (1 + p^{a-1})^i\alpha^i = (1 + p^{a-1}i + \cdots + p^{i(a-1)})\alpha^i.$$

Therefore $H^{2j}(\langle A\rangle, \mathbb{Z})^B$ is generated by

$$\begin{cases} \alpha_j = p\alpha^j & \text{if } p \nmid j \\ \alpha^j & \text{if } p \mid j \end{cases}.$$

In order to apply the periodicity isomorphism above it is necessary to calculate $\hat{H}^0$ and $\hat{H}^1$, where

$$\hat{H}^0(\langle B\rangle, H^{2j}(\langle A\rangle, \mathbb{Z}))$$

equals the invariant elements modulo the subgroup generated by the sum of powers of B, N_B. Note that

$$N_B(\alpha^j) = \frac{(1 + p^{a-1}j)^p - 1)}{p^{a-1}j}\cdot\alpha^j$$
$$= p \text{ (some multiple of } \alpha^j\text{), and } (p, j) = 1,$$

so that $N_B(\alpha^j)$ is an element of order p^{a-1}. Therefore the kernel of the norm map N_B has order p, and is generated by $p^{a-1}\alpha^j$. Allowing also for the case when p divides j:

$$\hat{H}^0(\langle B\rangle, H^{2j}(\langle A\rangle, \mathbb{Z})) = \begin{cases} 0 & \text{if } p \nmid j \\ \mathbb{Z}/p^{\alpha j} & \text{if } p \mid j \end{cases}.$$

Similarly

$$\hat{H}^1(\langle B\rangle, H^{2j}(\langle A\rangle, \mathbb{Z})) \cong \begin{cases} 0 & \text{if } p \nmid j \\ \mathbb{Z}/p & \text{if } p \mid j \end{cases},$$

where in both cases one takes the kernel of N_B modulo the image of the augmentation ideal.

These calculations determine the $E_2^{i,j}$ page of the spectral sequence; note that each term off the fibre has order zero or p, that $|E_2^{0,0}| = \infty$ and that $|E_2^{0,2j}|$ equals p^{a-1} or p^a, depending on whether p is a factor of j or not. Write $\zeta = \alpha^p$ as a generator for $E_2^{0,2p}$ and χ as a generator for $E_2^{1,2p}$. Then the various summands may be described as follows:

$E_2^{i,j} = 0$, if either j is odd, or $i > 0$ and $2p \nmid j$,
$E_2^{0,2j}$ is generated by $\alpha_{j_0}\zeta^l$, $j = j_0 + lp$, $0 \leqslant j_0 < p$,
$E_2^{2i,0}$ is generated by β^i, where, as in the definition of α, $\beta(B) = 1/p$,
$E_2^{2i,2pj}$ is generated by $\beta^i\zeta^j$, $i \geqslant 1, j > 0$, and
$E_2^{2i+1,2pj}$ is generated by $\chi\beta^i\zeta^{j-1}$, $i \geqslant 0, j > 0$.

Each of these generators is a universal cycle. For β^i this follows from the fact that the extension is split, and hence

$$\mathrm{Inf}: H^*(\langle B\rangle, \mathbb{Z}) \to H^*(G, \mathbb{Z})$$

is a monomorphism. The element α_1 survives to infinity, because it corresponds to a character in $\mathrm{Hom}(G, \mathbb{Q}/\mathbb{Z})$. Given the product structure on the columns of the E_2-page in the spectral sequence, the same holds for $\alpha_2, \ldots, \alpha_{p-1}$. The elements ζ and χ on the $(i, 2p)$-line survive because they cannot be hit from behind, and cannot themselves hit anything in front by the assertions just made for α_j and β^i. Hence, even as graded rings $E_2^{*,*} = E_\infty^{*,*}$, that is the spectral sequence collapses.

In dimension $2p$ ζ corresponds to an element of order p^a, rather than to one of order p^{a+1}, since by Theorem 3.10 the cohomology of G is not periodic. Furthermore, since χ represents a class in $H^{2p+1}(G, \mathbb{Z})$, by the skew commutativity of the product, $\chi^2 = 0$.

With a slight abuse of notation we now have the following table:

Generator	β	α_j	ζ	χ
Dimension	2	$2j, 1 \leqslant j < p$	$2p$	$2p+1$
Additive order	p	p^{a-1}	p^a	p

In a later chapter each even-dimensional generator will reappear as a characteristic class. For the moment note that the edge homomorphism on the fibre corresponds to restriction, and that

$$i_{G \to \langle A\rangle}\alpha_j = p\alpha^j,$$

an element of order p^{a-1} in $H^{2j}(\langle A\rangle, \mathbb{Z})$. Indeed from Chapter 3

$$i_{*\langle A\rangle \to G}\, i^*_{G \to \langle A\rangle}$$

equals multiplication by p, and

$$i^*_{G\to\langle A\rangle} i_{*\langle A\rangle\to G}(\alpha^j) = N_B\alpha^j.$$

So, we may, if necessary replace α_j by $\mathrm{Cor}(\alpha^j)$, $j=1, 2,\ldots,p-1$, see the description of $N_B\alpha^j$ above. With these identifications the relations between the ring generators in the table become clear.

Relations between the α_js or between the α_js and ζ follow by restriction to the fibre, and β generates a $(\mathbb{Z}/p)$-polynomial subring. Given the vertical periodicity defined by taking products with ζ (regarded as an element of $H^{2p}(\langle A\rangle, \mathbb{Z})$, the generators ζ and χ are multiplicatively independent. There remains the possibility of relations between elements obtained by corestriction from $\langle A\rangle$ and those obtained by inflation from $\langle B\rangle$. There can be none such, since by the reciprocity formula 2.6

(i) $\beta(i_*\alpha^j) = i_*(i^*\beta\cdot\alpha^j) = 0,$

the element β clearly vanishing on restriction to the subgroup generated by A.

Furthermore

(ii) If $\chi(i_*a^j)\neq 0$, then $\beta\chi(i_*\alpha^j)\neq 0$ by the periodicity of the cohomology of $\langle B\rangle$ with arbitrary coefficients, contradicting (i).

Summarising the calculations of the last few pages we have:

Theorem 4.5

If G is a non-abelian group of order p^{a+1}, which contains a cyclic normal subgroup of index p, then the ring $H^*(G, \mathbb{Z})$ is generated by

$$\{\beta, \alpha_1, \alpha_2, \ldots, \alpha_{p-1}, \zeta, \chi\}$$

subject to the multiplicative relations

$$\chi^2 = \alpha_j\chi = \alpha_j\beta = 0,$$

$$\alpha_i\alpha_j = p\alpha_{i+j}, p^2\zeta \quad \text{or} \quad p\zeta\alpha_{j_0},$$

depending on whether $i+j<p$, $i+j=p$ or $i+j=p+j_0$. The additive structure may be read off from the table above.

It is possible to carry out a similar calculation for a more general split metacyclic p-group. Indeed C.T.C. Wall [Wa] has shown that the spectral sequence of the extension always satisfies the collapsing condition $E_2^{i,j} = E_\infty^{i,j}$, and has used this to determine the additive structure of $H^*(G, \mathbb{Z})$. Special considerations then allow one to determine at least some of the ring structure – this is most effectively done in terms of Chern classes, see Chapter 6 below. For example, if the generator B has order p^b and the conjugacy condition is $A^B = A^r$, then in dimension $2j$ the fibre contributes a summand of order equal to $((r^j - 1), p^a)$.

Exercise

Assuming that the spectral sequence does collapse, carry out the analysis of Theorem 4.5 for the group

$$G = \{A, B \colon A^{p^a} = B^{p^2} = 1, A^B = A^r\}$$

in the cases (i) $r = p^{a-1} + 1$ and (ii) $a \geqslant 3, r = p^{a-2} + 1$.

As always the theory at the prime 2 has a special flavour. For example consider the dihedral group of order 2^{a+1} with presentation

$$D_{2^{a+1}} = \{A, B \colon A^{2^a} = B^2 = 1, A^B = A^{-1}\}$$

Either as above by hand or by appealing to the theorem of Wall just quoted one shows that the spectral sequence of the defining extension collapses. Note that because of the relation $A^B = A^{-1}$, the element $\alpha^2 \in H^4(\langle A \rangle, \mathbb{Z})$ is invariant and survives to infinity. Hence it is only necessary to examine the horizontal strip

$$\{E_2^{i,j} \colon 0 \leqslant j \leqslant 4\}$$

in order to calculate the integral cohomology. One obtains

Theorem 4.6

The ring $H^*(D_{2^{a+1}}, \mathbb{Z})$ *is generated over the integers by* $\{\alpha, \beta, \nu, \zeta\}$, *where* $\dim \alpha = \dim \beta = 2$, $\dim \nu = 3$, $\dim \zeta = 4$, *subject to the relations*

$$2^a\zeta = 2\nu = 2\alpha = 2\beta = 0;$$
$$\nu^2 = \beta\zeta, \alpha^2 = \alpha\beta.$$

The structure as a graded abelian group can be read off from the main result in [Wa]. For $a = 2$, the calculation is due to L. Evens, but see also [Le].

Notes and references

The general theory of the spectral sequence of the extension of one group by another is due to R. Lyndon and also to G. Hochschild & J.P. Serre. We have followed the presentation given in [Mac]; for an alternative rather more abstract approach, see the set of notes by S. Lang [La]. For both finite and infinite groups these spectral sequences have proved to be effective tools in the calculation of the cohomology ring.

In Chapter 1, see the section on low-dimensional interpretation, we saw that any group extension $K \rightarrowtail G \twoheadrightarrow G/K$, with K abelian, is determined by a homomorphism $G/K \to \mathrm{Aut}(K)$ and by an element $\upsilon \in H^2(G/K, K)$, which vanishes when the extension splits. The connection between υ and the differential d_2 is an interesting one, see [C–V]. The calculations in the second part of the chapter are based on §5 of [Le] and on the early paper of Wall, already mentioned. Both papers are extremely interesting and deserve to be more widely known. For another calculation using the spectral sequence with $\mathbb{F}_p$ rather than $\mathbb{Z}$ coefficients, see [N] and Theorem 9.4 below. The group considered is a p-Sylow subgroup of the symmetric group S_{p^n}, and the spectral sequence collapses for $\mathbb{F}_p$ but not for $\mathbb{Z}$. Later we shall have to consider such a non-collapsing example in the proof of Theorem 8.6 on p-groups of rank 2.

Problems

1. Obtain the low-dimensional 'inflation–restriction exact sequence' by directly checking exactness at each point. Note that if A is a not necessarily abelian group admitting a G-action, it is still possible to define $H^0(G, A)$ as the invariant elements, and $H^1(G, A)$ as a pointed set of equivalence classes of 'crossed maps $G \to A$', see J.P. Serre, *Corps locaux*, page 130. Examine the sequence of pointed sets

$$\to H^1(G/K, A^K) \xrightarrow[\mathrm{Inf}]{} H^1(G, A) \xrightarrow[\mathrm{Res}]{} H^1(K, A)$$

in this more general setting.

2. The most general metacyclic group of order p^{n+m} has a presentation

$$\{A, B: A^{p^n} = 1, B^{p^m} = A^{p^r}, A^B = A^k, r \geqslant 0,$$
$$k^{p^m} \equiv 1 \pmod{p^n}, p^r(k-1) \equiv 0 \pmod{p^n}\}$$

With the subgroup generated by A as normal subgroup examine the spectral sequence of the extension (with coefficients $\mathbb{Z}$) in the non-split case, i.e. with $r > 0$. Determine in particular whether the spectral sequence collapses ($E_2^{s,t} = E_\infty^{s,t}$), and identify the universal cycles on the fibre. Harder, but see [C-V] – what can you say about the image under inflation of $H^*(\mathbb{Z}^B/p^m, \mathbb{Z})$?

The reader may prefer to come back to this question after reading Chapters 5 and 8.

5

Representations and vector bundles

We now collect together the results needed from the theory of representations of a finite group over a field (which will usually be the complex numbers $\mathbb{C}$) and from the theory of fibre bundles. Since this is a book on group cohomology, we give no more than sketch proofs. However, for the reader unfamiliar with the material, we give, following each quoted theorem, a reference to a book or article, in which a detailed proof may be found.

Representations

A finite dimensional representation of the finite group G is a homomorphism $\rho: G \to \mathrm{GL}(n, \mathfrak{k})$, the general linear group of $n \times n$ invertible matrices over the field $\mathfrak{k}$, which will usually be algebraically closed and of characteristic zero. Indeed there will be little loss of generality in assuming $\mathfrak{k} = \mathbb{C}$, except in the very last chapter. By choice of a basis identify an n-dimensional vector space over $\mathfrak{k}$ with the space of n-tuples $\mathfrak{k}^n$, which using ρ can be given the structure of a (left) $\mathfrak{k}G$-module. Examples of such homomorphisms are (i) the trivial representation, $\rho(G) = 1_n$, and (ii) the regular representation of dimension equal to $[G:1]$, obtained by allowing G to act on $\mathfrak{k}G$ on the left. It is a classical result of Maschke that if the characteristic of $\mathfrak{k}$ does not divide the order of G, for example if $\mathfrak{k} = \mathbb{C}$, then the ring $\mathfrak{k}G$ is semisimple. Thus any submodule N of the representation module M is actually a $\mathfrak{k}G$-direct summand, so that M decomposes as

$$M = \coprod_{i \in I} M_i,$$

when each M_i is irreducible. This means that the only G-invariant subspaces of the finite dimensional vector spaces M_i are $\{0\}$ and M_i itself.

The set of isomorphism classes of finite dimensional G-modules admits an addition (direct sum) and a multiplication (tensor product over $\mathfrak{k}$), that is, we have a commutative monoid with multiplication, written $R(G)$ or $R_{\mathfrak{k}}G$, if it is important to specify the ground field. By Proposition 5.2 below $R(G)$ is actually a ring with subtraction formally defined by $-[M]=[-M]$. Now take $\mathfrak{k}=\mathbb{C}$.

Definition

The character χ_ρ of the representation ρ is the complex valued function $\chi_\rho: G\to\mathbb{C}$ given by $\chi_\rho(g)=\text{trace}(\rho(g))$.

Note that this definition is independent of the basis chosen for the vector space underlying the representation module M, also that $\chi_\rho=\rho$ if ρ is one-dimensional.

Proposition 5.1

With the notation already established, and with $\chi=\chi_\rho$

(i) $\chi(1)\in\mathbb{N}$, *and equals the dimension over* $\mathbb{C}$ of M,

(ii) $\chi(g^{-1})=\overline{\chi(g)}$

(iii) χ *is well-defined on the family* $\{C_j: j=1,2,\ldots,h\}$ *of conjugacy classes of elements of* G.

(iv) $(\chi_1+\chi_2)(g)=\chi_1(g)+\chi_2(g)$ *and* $\chi_1\chi_2(g)=\chi_1(g)\chi_2(g)$.

Proof. Immediate from the definitions.

Any function $\phi: G\to\mathbb{C}$ satisfying conditions (i)–(iv) is called a *class function*, denote the set of all such by $\mathfrak{X}(G)$. The value $\chi_\rho(g)$ of χ_ρ at g is necessarily an algebraic integer; if m is a multiple of the exponent of the finite group G, then $(\rho(g))^m=1_n$, from which it follows that trace$(\rho(g))$ is a sum of mth roots of unity.

Define an inner product on the space $\mathfrak{X}(G)$ of class functions by

$$\langle\phi,\psi\rangle=\frac{1}{[G:1]}\sum_{g\in G}\overline{\phi(g)}\cdot\psi(g).$$

Proposition 5.2

If χ_1 *and* χ_2 *are the characters of two non-isomorphic irreducible complex representations, then* $\langle\chi_i,\chi_j\rangle=\delta_{ij}$.

It follows from 5.2 that the representation module M contains a well-defined multiple n_i of copies of the irreducible module M_i. Furthermore M and N are isomorphic over the ring G if and only if $\chi_M=\chi_N$, the inner product $\langle \chi_M, \chi_M \rangle$ is integral, and M is irreducible if and only if $\langle \chi_M, \chi_M \rangle = 1$. As an abelian group $R_{\mathbb{C}}(G)$ is free on a family of generators indexed by the irreducible representations (or by their characters), which justifies the simple-minded definition of subtraction given above. An element of the character ring is an integral combination $\coprod_{i \in I} n_i \chi_i$ of irreducible characters, called a virtual character. Such a virtual character is the formal difference of the characters of two homomorphisms of G into $\mathrm{GL}(\cdot, \mathbb{C})$; thus a virtual character is a character if $n_i \geqslant 0$ for all i. If χ_{reg} denotes the character of the regular representation of G, then

$$\chi_{\mathrm{reg}}(1) = [G{:}1],$$

and

$$\chi_{\mathrm{reg}}(g) = 0, \text{ if } g \neq 1.$$

Proposition 5.3

(i) *The number of distinct irreducible complex representations of G equals h, the number of conjugacy classes of elements.*

(ii) *If the irreducible representation M_i has dimension n_i over $\mathbb{C}$, then*

$$\sum_{i=1}^{h} n_i^2 = [G{:}1].$$

(iii) *The dimension n_i divides the order of G.*

Sketch proof, for the details see [Se 2, §2.4, §2.25, §6.4]

(i) The characters of the irreducible representations form a normal orthogonal basis for the space of class functions $\mathfrak{X}(G)$.

(ii) Decompose the character of the regular representation as the sum

$$\chi_{\mathrm{reg}} = \sum_{i=1}^{h} n_i \chi_i,$$

and apply the inner product on both sides.

(iii) This depends on the subsidiary proposition, see Serre [Se 2, Proposition 10, §6.3] that $(1/n_i)(\sum_{g \in C_j} \chi(g))$ is an algebraic integer. More generally one can prove that if A is a maximal, normal abelian subgroup of G, then n_i divides the index $[G{:}A]$.

Let $i: K \to G$ be the inclusion homomorphism of the subgroup K. As in the case of cohomology there are change of group homomorphisms

$$i^!: R(G) \to R(K),$$

$$i_!: R(K) \to R(G),$$

$$c_g: R(K) \to R(K^g), g \in G.$$

The ring homomorphism $i^!$ is the natural forgetful map, which restricts the action on the module M to the subgroup K. The map $i_!$ is called induction or transfer, and may be defined either on modules or on their characters. For the former – compare the definition of corestriction in cohomology,

$$i_!M = \mathbb{C}G \bigotimes_K M,$$

where $\mathbb{C}G$ is considered as having a left G- and a right K-structure.

In terms of characters,

$$i_!\chi(g) = \frac{1}{[K:1]}\left(\sum_{\substack{x\in G,\\ g^x\in K}} \chi(g^x) \right).$$

The third change of group map c_g is induced by conjugation by the group element g, thus

$$(c_g\chi)(x^g) = \chi(x).$$

If K is normal in G, c_g is a ring endomorphism; if $K = G$, c_g is trivial (compare once more with cohomology). Transfer is a homomorphism of $R(G)$-modules; this follows from either of the Frobenius reciprocity formulae:

$$i_!(i^!M \otimes N) = M \otimes i_!N \text{ or } \langle i^!\phi, \psi\rangle_K = \langle \phi, i_!\psi\rangle_G.$$

Definition

K is an elementary subgroup of G if

$$K = C_r \times P,$$

where C_r is cyclic and P is a p-subgroup of coprime order, that is $p \nmid r$.

Theorem 5.4 (R. Brauer)

Let $\phi \in \mathfrak{X}(G)$ be a complex valued class function. Then ϕ is a virtual character if and only if the restriction of ϕ to each elementary subgroup K of G is a virtual character.

Proof. See [Sw1, Section 4], or [Se 2, Theorem 22, §§10–11].

One can state the conclusion of Brauer's Theorem in terms of an exact sequence, see [At]. Define homomorphisms $\Psi'_{r,s,t}\colon R(K_r) \to R(K_s \cap K_t)$ and $\Psi''_{r,u,g}\colon R(K_r) \to R(K_u^g)$ by

$$\Psi'_{r,s,t} = \begin{cases} i^!, & r = s, r \neq t, \\ -i^!, & r \neq s, r = t, \\ 0, & \text{otherwise} \end{cases}$$

and

$$\Psi''_{r,u,g,} = \begin{cases} 1, & K_r = K^g_u, u \neq r, \\ -c_g, & K_r \neq K^g_u, u = r, \\ 1 - c_g, & K_r = K_u = K^g_u, \\ 0, & \text{otherwise.} \end{cases}$$

Here r, s, t, u all index the family of elementary subgroups of G, g indexes the elements of G, and $\Psi = (\Psi'_{r,s,t}, \Psi''_{r,u,g})$. The components of Ψ have been chosen so that for a family of class functions $\{\phi_r\}$ defined on the elementary subgroups of G and lying in the kernel of Ψ

$$i^!_{K_s \to K_s \cap K_t}(\phi_s) = i^!_{K_t \to K_s \cap K_t}(\phi_t), \quad \text{and} \quad \phi^g_u = \phi_{K^g_u},$$

for all classes of s, t and g. The following sequence of representation rings is then exact:

$$0 \to R(G) \xrightarrow[i^!]{} \coprod_r R(K_r) \xrightarrow[\Psi]{} \coprod_{s,t} R(K_s \cap K_t) + \coprod_{u,g} R(K^g_u).$$

The reader should compare this sequence with the sequence of cohomology groups obtained in Theorem 3.3. By fixing attention on a single elementary subgroup K_r, in particular $K_r = G_p$ a p-subgroup of maximal order, we are able to describe those representations of the subgroup which are extendable in some way (in general non-uniquely) to a representation of G.

For supersolvable groups, a class which contains groups of prime power order, there is the following useful result, which can be thought of as a weak splitting principle for representations (see Proposition 5.9 in the next section).

Theorem 5.5 (Blichfeldt)

Let ρ be an irreducible representation of the supersolvable group G of dimension greater than 1. Then $\rho = i_!\sigma$ for some one-dimensional representation σ of some proper subgroup K of G.

Proof. See [Se 2, §9.5].

In this subsection we collect together various facts about the relation between the structure of $R(G)$ as a λ-ring (taking into account the exterior powers over $\mathbb{C}$ of a representation module M) and the field of definition of a character χ. Note first that an algebraic automorphism σ of the complex numbers induces an automorphism σ^* of $\mathrm{GL}(n, \mathbb{C})$ by allowing σ to act on the matrix entries. The automorphism σ also induces a map on $R(G)$, and the image ρ^σ of a representation under σ_* has character $\chi_{\rho^\sigma} = \sigma\chi_\rho$. In this way a subgroup $\mathfrak{G}$ of the Galois group of $\mathbb{C}$ over $\mathbb{Q}$ can be made to act on $R(G)$. Let m be a multiple of the exponent of G, $\mathfrak{k}$ some subfield of $\mathbb{C}$, and write $\mathfrak{k}_m = \mathfrak{k}(\zeta_m)$, the mth cyclotomic extension of $\mathfrak{k}$, where ζ_m is a primitive mth root of unity in $\mathbb{C}$. The group $\mathrm{Gal}(\mathfrak{k}(\zeta_m)/\mathfrak{k})$ is finite abelian.

Each character χ of the group G takes values in $\mathfrak{k}_m$, so the action of $\mathrm{Gal}(\mathbb{C}/\mathfrak{k})$ factors through that of its homomorphic image under restriction $\mathrm{Gal}(\mathfrak{k}_m/\mathfrak{k})$. Furthermore a necessary and sufficient condition for χ to take values in $\mathfrak{k}$ itself is that $\chi^\sigma = \chi$ for all $\sigma \in \mathrm{Gal}(\mathbb{C}/\mathfrak{k})$.

The representation ring $R(G)$ may be given the structure of a λ-ring, see [A–T], the exterior power $\Lambda^k_{\mathbb{C}} M$ becoming a G-module by composition of ρ with the usual representation of $\mathrm{Aut}_{\mathbb{C}} M$ on $\mathrm{Aut}_{\mathbb{C}} \Lambda^k M$. With the conventions that

$$\Lambda^0 M = 1, \quad \Lambda^1 M = M$$

and the rule that

$$\Lambda^k(M_1 \otimes M_2) = \coprod_{i=0}^{k} (\Lambda^i M_1 \otimes \Lambda^{k-i} M_2),$$

the map

$$M \longmapsto 1 + Mt + \Lambda^2 M t^2 + \cdots$$

is a homomorphism from the additive semi group of isomorphism classes of $\mathbb{C}G$-modules into $1 + R(G)[[t]]$, the multiplicative group of formal power series with constant term 1. Now define the Adams operations $\{\psi^k\}$ by the rule

$$\psi^k(M) = N_k(\Lambda^1 M, \ldots, \Lambda^k M),$$

where N_k is the kth Newton polynomial, that is, the polynomial expressing $x_1^k + \cdots + x_k^k$ in terms of the first k elementary symmetric functions

$$y_i(x_1, \ldots, x_k) \quad 1 \leqslant i \leqslant k.$$

The polynomial N_k may be defined recursively by the formula

$$N_k - y_1 N_{k-1} + \cdots + (-1)^k k y_k = 0;$$

in particular $N_1 = y_1$, $N_2 = y_1^2 - 2y_2$.

By direct calculation we obtain

Proposition 5.6

(i) $(\chi_{\Lambda^k\rho})(g) = y_k(\omega_1, \ldots, \omega_n)$, *where* $\omega_1, \ldots, \omega_n$ *are the eigenvalues of the matrix* $\rho(g)$

(ii) $(\psi_{\psi}{}^{k}{}_{\rho})(g) = \omega_1^k + \cdots + \omega_n^k = \chi_\rho(g^k)$.

In particular $\psi^k = \psi^{k+m}$ for all values of k.

(iii) *The operation* $\chi_\rho \longmapsto \chi_{\psi^k\rho} = \psi^k\chi_\rho$ *is compatible with direct sums and tensor products, so that*

$$\psi^k : R(G) \to R(G)$$

is a ring homomorphism.

The next proposition follows from the preceding discussion.

Proposition 5.7

Let m be a multiple of the exponent of the finite group G. Given $\sigma \in \operatorname{Aut} \mathbb{C}$, *choose a positive integer k such that* $\sigma\omega = \omega^k$ *for all mth roots of unity* ω. *Then for all* $\chi \in R(G)$

$$\chi^\sigma = \psi^k \chi.$$

The highest common factor $(k, m) = 1$, because σ is an automorphism, so k has a well-defined congruence class $[k]$ modulo m. Each such class in $\mathbb{Z}/m^\times$ is associated with some field automorphism; and for a finite extension field $\mathfrak{k}$ of $\mathbb{Q}$, the Galois group $(\mathfrak{k}_m/\mathfrak{k})$ determines a subgroup of the group of units $\mathbb{Z}/m^\times$ according to the diagram

$$\begin{array}{ccc} \operatorname{Gal}(\mathbb{C}/\mathfrak{k}) & \xrightarrow{\mu_{\mathfrak{k},m}} & \mathbb{Z}/m^\times \\ & \searrow & \uparrow \lambda_{\mathfrak{k},m} \\ & & \operatorname{Gal}(\mathfrak{k}_m/\mathfrak{k}) \end{array}$$

Here $\mu_{k,m}(\sigma) = [k]$ if and only if $\sigma\omega = \omega^k$ for all mth roots of unity. Furthermore

Proposition 5.8

If m is a multiple of the exponent of G and $\mathfrak{k}$ *is a finite extension of the rational numbers* $\mathbb{Q}$, *then* χ_ρ *takes values in* $\mathfrak{k}$ *if and only if* $\psi^k\chi_\rho = \chi_\rho$ *for all positive values of k, such that the congruence class* $[k]$ *lies in the image of* $\lambda_{\mathfrak{k},m}$.

Coordinate bundles

Intuitively a fibre bundle ξ consists of a total space E, a continuous projection map $p: E \twoheadrightarrow X$, and a local product structure, that is, each point x of the base space X belongs to an open subset U such that $p^{-1}U$ is homeomorphic to $U \times F$. The fixed space F is called the fibre – note that as a set $E = \bigcup_{x \in X} p^{-1}(x)$, each inverse image being homeomorphic to F. The problem is to make formal the relation between overlapping local trivialisations – we do this by means of a coordinate bundle, starting with the data

$$\{X, \{U_i : i \in I\}, \Gamma, F, \gamma_{ji}\}.$$

Here the open sets U_i form a covering of the topological space X, Γ is a topological group acting on F, and $\gamma_{ji}: U_i \cap U_j \to \Gamma$ is a family of continuous functions satisfying

(1) $\gamma_{ii} = 1$,
(2) $\gamma_{ij} = \gamma_{ji}^{-1}$

and

(3) $\gamma_{ki} = \gamma_{kj}\gamma_{jk}$

on the domain $U_i \cap U_j \cap U_k$. Without loss of generality we suppose that as a group of transformations Γ is effective in the sense that $\gamma f = f$ for all f in F only if γ is the identity. Define the total space E to be the quotient space

$$\bigcup_{i \in I} (U_i \times F \times i)/R,$$

where $(x, f, i)R(x', f', j)$ provided that $x = x'$ and $\gamma_{ji}(x)(f') = f$. Thus E is obtained from the disjoint union of the product spaces $U_i \times F$ by identifying copies of $x \times F$ by means of the action of Γ when $x \in U_i \cap U_j$. The projection map p takes the equivalence class of (x, f, i) to x, and is continuous by the definition of the quotient topology. The local product structure is clear from the definition, and the maps γ_{ji} are called the coordinate transformations. We summarise the defining data by means of the Greek letter ξ. In order to make the definition independent of a particular open covering of X we can either work with equivalence classes of coordinate bundles over X, the families γ_{ji} and γ'_{rs} being compatible in an obvious sense, or suppose that (U_i, γ_{ji}) can be embedded in some maximal family of open subsets and coordinate transformations. In practice though one always works with a representative coordinate bundle from the class. A morphism of (coordinate) bundles is given by a commutative diagram

$$\begin{array}{ccc} E_1 & \xrightarrow{\tilde{f}} & E_2 \\ {\scriptstyle p_1}\downarrow & & {\scriptstyle p_2}\downarrow \\ X_1 & \xrightarrow[f]{} & X_2 \end{array}$$

such that $\tilde{f}$ induces a homeomorphism on each fibre, compatible with the action of Γ.

Examples

(1) $F = \Gamma$ and Γ acts on itself by left translation. In this case the bundle is said to be principal; in the next chapter Γ will be discrete. Any bundle defined using the same coordinate transformations, but allowing Γ to act on some topological space F as fibre is said to be *associated* to the principal Γ-bundle.

(2) $F = \mathbb{C}^n$ and Γ equals some subgroup of $GL(n, \mathbb{C})$. In this case E is the total space of a *vector bundle*; the set of equivalence classes of all such over X will be denoted $\text{Vect}_n X$. A morphism of vector bundles induces a linear isomorphism on each fibre. Familiar examples from differential geometry are the tangent bundle to a C^∞-manifold and its associated tensor bundles.

There are two particularly important constructions for fibre bundles:

1. *Induced bundles*

Let ξ be a bundle over X_2 with group Γ and fibre F, and let $f: X_1 \to X_2$ be a continuous map. The data

$$f^!\xi = \{X_1, \{f^{-1}U_i\}, \Gamma, F, \gamma_{ji} \circ f\}$$

defines the induced (coordinate) bundle, the total space of which consists of all pairs (e_2, x_1) such that $p_2 e_2 = f x_1$.

2. *Bundles defined by a representation*

Let ξ be a principal bundle over X with group Γ_1, let $\phi: \Gamma_1 \to \Gamma_2$ be a continuous homomorphism, and let Γ_2 act on F_2. Then with the possibility of varying the fibre the data

$$\xi\phi_\natural = \{X, \{U_i\}, \Gamma_2, F_2, \phi \circ \gamma_{ji}\}$$

defines a new coordinate bundle.

By means of these constructions one extends the operations of linear algebra to the family $\{\text{Vect}_n X : n \geqslant 0\}$. At the intuitive level one forms the sum, tensor product or exterior power of $\mathbb{C}$-vector spaces over each point of X, and then defines the total space to be the union of the fibres. Formally one needs to be precise about the coordinate transformations, as an example consider the sum. Let $\Delta: X \to X \times X$ be the diagonal map $\Delta(x) = (x, x)$, and ξ_i a vector bundle of fibre dimension equal to n_i ($i = 1$, 2). There is a natural coordinate bundle $\xi_1 \times \xi_2$ defined over $X \times X$ with structural group contained in $GL(n_1,\mathbb{C})\times GL(n_2,\mathbb{C})$ and coordinate transformations $(\gamma_{ji}^{(1)},\gamma_{rs}^{(2)})$.

Write

$$\oplus: GL(n_1,\mathbb{C})\times GL(n_2,\mathbb{C})\to GL(n_1+n_2,\mathbb{C})$$

for the inclusion representation.

Definition

The Whitney sum $\xi_1\oplus\xi_2$ equals $\Delta^!(\xi_1\times\xi_2)\oplus_!$ with fibre $\mathbb{C}^{n_1+n_2}$.

Similarly one defines $\xi_1\otimes\xi_2$ (of fibre dimension n_1n_2) and the exterior power coordinate bundles $\Lambda^i\xi$, $0\leqslant i\leqslant n$. The basic constructions (1) and (2) above are compatible with these algebraic operations; $\oplus$ is commutative and associative, $\otimes$ is commutative, associative and distributive with respect to $\oplus$, and f commutes with $\oplus$, $\otimes$ and Λ^i.

Definition

The vector bundle η is called a complex line bundle if its structural group is a subgroup of $\mathbb{C}^\times = GL(1,\mathbb{C})$. The tensor product $\otimes$ defines a commutative, associative multiplication on $\mathrm{Vect}_1^{\mathbb{C}}(X)$.

The importance of line bundles in the construction of characteristic classes is shown by the following 'splitting principle':

Proposition 5.9

Given a complex vector bundle ξ over X, there exists a continuous map $h: X^1\to X$ such that

(i) $h^*: H^*(X,\mathbb{Z})\to H^*(X^1,\mathbb{Z})$ *is a monomorphism, and*

(ii) $h^!\xi\cong\eta_1\oplus\cdots\oplus\eta_n$ *is an n-fold sum of line bundles.*

Proof. See [Hs, page 235].

If one compares Propositions 5.9 and 5.5 one sees that the common idea is to lift the bundle or representation to some space or group, over which it splits into a sum of 1-dimensional objects. The difference between representations of a supersolvable group and vector bundles is that in the latter case one obtains an associated monomorphism in cohomology. There are examples to show that, at least with coefficients in $\mathbb{Z}$, this cannot be achieved for representations.

Recall that the complex projective space $\mathbb{C}P(n)$ is defined by the set of 1-

dimensional subspaces of $\mathbb{C}^{n+1}$. Topologically the assignment of homogeneous coordinates

$$(z_0, \ldots, z_n) \longmapsto [z_0, \ldots, z_n], \quad \text{with} \sum_{j=0}^{n} z_j \bar{z}_j = 1,$$

defines the projection map of an S^1-bundle

$$S^{2n+1} \longrightarrow \mathbb{C}P(n).$$

Here $S^1 = U(1) \subseteq \mathrm{GL}(1, \mathbb{C})$. Inside the complement of the hyperplane $z_j = 0$, assign the inhomogeneous coordinates $(z_0/z_j, \ldots, \hat{j}, \ldots, z_n/z_j)$, $0 \leqslant j \leqslant n$. Over each complex projective space is defined the associated line bundle, γ_n the total space of which is contained in $\mathbb{C}P(n) \times \mathbb{C}^{n+1}$,

$$E = \{(L, \mathbf{z}) : \mathbf{z} \in L\}.$$

If $\mathbb{C}P(\infty) = \bigcup_n \mathbb{C}P(n)$ the line bundle γ_∞ is defined formally in the same way; write $\gamma = \gamma_\infty$ and note that γ_{n+1} restricts to γ_n under the inclusion of $\mathbb{C}P(n)$ in $\mathbb{C}P(n+1)$.

The integral cohomology ring $H^*(\mathbb{C}P(\infty); \mathbb{Z})$ is well known to be a polynomial ring over $\mathbb{Z}$ on a single two-dimensional generator, which may be taken to be the dual of the class carried by $S^2 = \mathbb{C}P(1)$, oriented by the complex structure. The additive structure is clear from the fact that $\mathbb{C}P(n+1)$ is obtained from $\mathbb{C}P(n)$ by adjoining a single cell of dimension $2(n+1)$; the multiplicative structure may be verified by considering the intersection properties of the geometric carriers of dual homology classes. Furthermore, since

$$S^\infty = \bigcup_n S^{2n+1}$$

is a contractible space, $\mathbb{C}P(\infty)$ is a space of type $(\mathbb{Z}, 2)$, that is

$$\pi_2(\mathbb{C}P(\infty)) \cong Z \quad \text{and} \quad \pi_i(\mathbb{C}P(\infty)) = 0, \text{ otherwise.}$$

These facts may be found in any good book on algebraic topology, for example that of E.H. Spanier.

Proposition 5.10

If X is a CW-complex the elements of the homotopy set $[X, \mathbb{C}P(\infty)]$ are in (1–1) correspondence both with $\mathrm{Vect}_1^{\mathbb{C}}(X)$ and with $H^2(X, \mathbb{Z})$. Since $\mathbb{C}P(\infty)$ has a unique homotopy multiplication, there is an algebraic isomorphism

$$c_1 : (\mathrm{Vect}_1^{\mathbb{C}}(X), \otimes) \xrightarrow{\simeq} H^2(X, \mathbb{Z})$$

We take this isomorphism as the definition of the (first) Chern class of a complex line bundle, and from now on we shall assume that $H^*(\mathbb{C}P(\infty), \mathbb{Z})$ is generated by $c_1(\gamma)$.

Outline of the proof. The identification of the second cohomology group $H^2(X, \mathbb{Z})$ with the set of homotopy classes $[X, \mathbb{C}P(\infty)]$ is again standard algebraic topology. The map

$$[X, \mathbb{C}P(\infty)] \to \mathrm{Vect}_1^{\mathbb{C}}(X)$$

sends the homotopy class of a map f to the induced bundle $f^{!}\gamma$. In order to see that this map is onto, one associates with the line bundle η a map $g: E(\eta) \to \mathbb{C}^m$, where m is large, possibly infinite. If X is a CW-complex g is constructed by first covering X by a countable, locally finite family of open subsets U_i, over each of which η is trivial. It is now easy to first map $E(\eta)|U_i$ into some subspace of $\mathbb{C}^m$ in such a way that g is a linear monomorphism on each fibre, and then, using the unboundedness of m, to piece the local maps together. Given g it is straightforward to construct a bundle map $\tilde{f}: E(\eta) \to E(\gamma)$, see [Hs] page 30 for the details. One–oneness follows from the fact that $f_1^{!}\gamma \cong f_2^{!}\gamma$ if and only if f_1 is homotopic to f_2, and the set isomorphism which now exists between $H^2(X, \mathbb{Z})$ and $\mathrm{Vect}_1^{\mathbb{C}}(X)$ is algebraic because of the homotopy uniqueness of multiplication on $\mathbb{C}P(\infty)$.

Given Propositions 5.9 and 5.10 it is now easy to motivate the construction of the Chern classes $c_k(\xi)$ in $H^{2k}(X, \mathbb{Z})$ for an arbitrary complex vector bundle over the CW-complex X, which need not be finite or even finite-dimensional.

Theorem 5.11

For each vector bundle ξ in $\mathrm{Vect}_n^{\mathbb{C}}(X)$ there is a unique family of integral cohomology classes $c_k(\xi)$ in $H^{2k}(X, \mathbb{Z})$, such that

(1) $c_k(\xi) = 0$, $k > n$,

(2) *if $f: X^1 \to X$ is continuous, then $c_k(f^{!}\xi) = f^* c_k(\xi)$,*

(3) *if $c.(\xi) = 1 + c_1(\xi) + \cdots + c_n(\xi)$, then $c.(\xi_1 \oplus \xi_2) = c.(\xi_1)c.(\xi_2)$*

and

(4) $c_1(\gamma) \in H^2(\mathbb{C}P(\infty), \mathbb{Z})$

coincides with the dual of the class carried by S^2, naturally orientated.

Rather than a formal proof, for which see [Hs, Chapter 16], we now make the existence of such Chern classes plausible in the light of the preceding discussion.

First let ξ be a line bundle, classified by a map $f: X \to \mathbb{C}P(\infty)$, and define

$$c.(\xi) = 1 + f^*(c_1(\gamma)).$$

If ξ is a sum of line bundles, $\xi = \eta_1 \oplus \cdots \oplus \eta_n$, define

$$\begin{aligned} c.(\xi) &= (1 + f_1^*(c_1\gamma))(1 + f_2^*(c_1\gamma)) \cdots (1 + f_n^*(c_1\gamma)) \\ &= c.(\eta_1)c.(\eta_2)\ldots c.(\eta_n). \end{aligned}$$

Clearly properties (1)–(4) hold for such decomposable bundles, and determine the components of the total Chern class $c.(\xi)$ uniquely. In the general case let $h: X^1 \to X$ be a splitting map for ξ, that is,

$$h^!\xi = \eta_1 \oplus \cdots \oplus \eta_n.$$

In order to define $c.(\xi)$ one needs to know that $c.(h^!\xi)$ lies in the image of $H^*(X, \mathbb{Z})$ in $H^*(X^1, \mathbb{Z})$ under h^*. That this is so follows either by a variant of the argument used in the proof of the splitting principle, see [Hs, pages 230–3], or by using the cohomology of the classifying space $\mathrm{BU}(n)$, see below. In either case the construction so far, together with the fact that h^* is a monomorphism, proves uniqueness.

In later chapters our particular interest will be the relation between the complex representation ring $R(G)$ and $H^*(G, \mathbb{Z})$ given by the Chern classes. However there are other families of characteristic classes defined for real as opposed to complex representations. Thus suppose that ξ is a real vector bundle with $F = \mathbb{R}^n$ and structural group Γ contained in $\mathrm{GL}(n, \mathbb{R})$. A real line bundle is classified by a map into $\mathbb{R}P(\infty) = \bigcup_n \mathbb{R}P(n)$, and $H^*(\mathbb{R}P(\infty), \mathbb{F}_2)$ is polynomial on a one-dimensional generator w (note the change of coefficients). As in Theorem 5.10 there exist unique classes $w_k(\xi)$ in $H^k(X, \mathbb{F}_2)$, called the Stiefel–Whitney classes, such that axioms (1)–(3) hold and

$$w.(\gamma^{\mathbb{R}}) = 1 + w.$$

The definition also implies that, where the equality makes sense, $c_1(\mathrm{mod}\ 2) = w_2$.

Using the Thom isomorphism, see [Hs, Chapter 16], one defines the Euler class $e(\xi) \in H^n(X, \mathbb{Z})$ for an oriented real vector bundle. 'Oriented' means that the coordinate transformations γ_{ji} for ξ may be chosen to take values in the component $\mathrm{SL}(n, \mathbb{R})$ of the structural group. The Euler class can also be used as a starting point in the construction of the Chern classes – define the top dimensional class $c_n(\xi)$ for $\xi \in \mathrm{Vect}_n^{\mathbb{C}}(X)$ to be the Euler class of the underlying real vector bundle in $\mathrm{Vect}_{2n}^{\mathbb{R}}(X)$, see [Mi].

Reversing this idea, let ξ be an oriented real vector bundle, and complexify its structural group by means of the inclusion $\mathrm{SL}(n, \mathbb{R})$ in $\mathrm{GL}(n, \mathbb{C})$. This defines a bundle $\xi \otimes \mathbb{C}$ in $\mathrm{Vect}_n^{\mathbb{C}}(X)$ by construction (2) above, for which we write

$$p_k(\xi) = (-1)^k c_{2k}(\xi \otimes \mathbb{C}) \in H^{4k}(X, \mathbb{Z}).$$

These classes are called the Pontrjagin classes; the odd indexed Chern classes are of little interest, since $\xi \otimes \mathbb{C}$ is self-conjugate.

Classifying spaces

There is an alternative approach to characteristic classes which uses the cohomology ring of the classifying space for the bundles concerned, see Chapter 18 in [Hs]. At least over a CW-complex X any complex bundle ξ is equivalent to $f^*\gamma(n)$, where $\gamma(n)$ is the universal bundle of fibre dimension n over the Grassmann manifold of n-subspaces in $\mathbb{C}^k$ (k-large), $G_{n,k}^{(\mathbb{C})}$. To avoid problems with the dimension of X take

$$G_{n,\infty}^{(\mathbb{C})} = \bigcup_k G_{n,k}^{(\mathbb{C})}.$$

As for the case $n = 1$ already considered the total space of $\gamma(n)$ consists of pairs $(L, \mathbf{z})$, where the n-vector $\mathbf{z}$ lies in the subspace L. In the light of this universal property one writes

$$G_{n,\infty}^{(\mathbb{C})} = \mathrm{BGL}(n, \mathbb{C});$$

Since $\mathbb{C}P(\infty) = \mathrm{BGL}(1, \mathbb{C})$, the inclusion of the n-torus in $\mathrm{GL}(n, \mathbb{C})$ induces a continuous map

$$j: \underbrace{\mathbb{C}P(\infty) \times \cdots \times \mathbb{C}P(\infty)}_{n} \to \mathrm{BGL}(n, \mathbb{C}).$$

Theorem 5.12

If $H^*(\mathbb{C}P(\infty)_i, \mathbb{Z}) = \mathbb{Z}[z_i]$, $1 \leqslant i \leqslant n$, *the induced map* j^*: $H^*(\mathrm{BGL}(n, \mathbb{C}), \mathbb{Z}) \to \mathbb{Z}[z_1, \ldots, z_n]$ *is a monomorphism the image of which is the polynomial algebra generated by the elementary symmetric functions* $y_i(z_1, \ldots, z_n)$, $1 \leqslant i \leqslant n$.

Call the classes $y_1, \ldots, y_n$ the universal Chern classes; if $\xi \in \mathrm{Vect}_n^{\mathbb{C}}(X)$ is classified by the map $f: X \to \mathrm{BGL}(n, \mathbb{C})$, define

$$c_k(\xi) = f^* y_k, \quad 1 \leqslant k \leqslant n.$$

This definition, augmented by $c_0(\xi) = 1$, satisfies properties (1)–(4) in

Theorem 5.10. The only non-trivial point is the verification of (3), which depends on using the bundle $\gamma(n_1 + n_2)$ to classify $\gamma(n_1) \times \gamma(n_2)$. Once this has been done property (4) shows that the two definitions of the Chern classes coincide. Note that the calculation in 5.11 closes the gap in the sketch proof given for 5.10 – with $h\colon X^1 \to X$ a splitting map for ξ, consider the diagram

$$\begin{array}{ccc} X^1 & \xrightarrow{f^1} & \mathbb{C}P(\infty) \times \cdots \times \mathbb{C}P(\infty) \\ \downarrow h & & \downarrow j \\ X & \xrightarrow{f} & \mathrm{BGL}(n, \mathbb{C}) \end{array}$$

which is such that $h^* f^* = f^{1*} j^*$. This is enough to show that

$$c.(h^!\xi) = c.(\eta_1 \oplus \cdots \oplus \eta_n) \in h^* H^*(X, \mathbb{Z}).$$

Notes and references

The best book to fill out the details for the first section is undoubtedly the introduction to representation theory by J.P. Serre [Se2]. It is harder to give a single source for the second section, although the omitted detailed proofs can be found in [Hs]. For a proof of Theorem 5.12 see the notes of J.F. Adams [Ad 2].

6

Bundles over the classifying space for a discrete group

Flat bundles

Let G be a discrete group, which for the time being may be infinite. A principal G-bundle over the topological space X is determined by an open covering $\{U_i : i \in I\}$ of X and a family of coordinate transformations $\gamma_{ji}: U_i \cap U_j \to G$. Since γ_{ji} is continuous and G has the discrete topology, each such function is locally constant.

Example

Let $X = S^1$, $U_0 = S^1 - \{1\}$, $U_1 = S^1 - \{-1\}$, so that $U_0 \cap U_1$ is the union of two disjoint open arcs. Define $\gamma_{01}: U_1 \cap U_0 \to \{\pm 1\}$ to take the value $+1$ on one component and -1 on the other. The real line bundle associated to the principal $(\mathbb{Z}/2)$-bundle over S^1 with this data using the natural inclusion of $\mathbb{Z}/2$ in $\mathrm{O}(2)$ has total space homeomorphic to an open Möbius band.

One method of constructing discrete G-bundles over the connected space X is to take an epimorphism $\phi: \pi_1 X \twoheadrightarrow G$, and write $\bar{X}$ for the covering space with fundamental group isomorphic to the kernel of ϕ, on which G acts freely as a group of covering transformations. There is a covering $\{U_i : i \in I\}$ of X, such that the inverse image of U_i in $\bar{X}$ is, up to homeomorphism, a disjoint union of copies of U_i indexed by the elements of G. Comparison of inverse images over U_i and U_j determines locally constant coordinate transformations $\gamma_{ji}: U_i \cap U_j \to G$, giving $\bar{X}$ the required structure.

If $\rho: G \to \mathrm{GL}(n, \mathbb{C})$ is a homomorphism, then by composing ρ with each of the maps γ_{ji}, and allowing $\mathrm{GL}(n, \mathbb{C})$ to act in the usual way on $\mathbb{C}^n$, we obtain a vector bundle over X, whose coordinate transformations are still locally constant. In particular each map $\gamma_{ji}: U_i \cap U_j \to \mathrm{GL}(n, \mathbb{C})$ factors through $GL(n,\mathbb{C})^\delta$, where the superscript δ denotes the discrete topology. Such a vector bundle is called a flat vector bundle over the space X. This construction, a special case of (2) on page 57 above, applies when $\pi_1 X \cong G$, the universal covering space of X is homotopic to a point, and ϕ is the identity. Such a space X exists for any discrete group G, as we now show using a topological construction analogous to the standard resolution in Chapter 1.

Recall that a generator in dimension k for the standard resolution is labelled by a $(k+1)$-tuple $(g_0,\dots,g_k)$ of elements from G. Take these as the vertices of a simplex $(t_0g_0,\dots,t_kg_k)$, where $0 \leqslant t_i \leqslant 1$ and $\sum_{i=0}^k t_i = 1$. Let EG be the quotient space obtained by identifying faces (some $t_i = 0$) according to the differential d in Chapter 1, and define an action of G on EG by allowing G to act on individual cells according to the rule

$$g(t_0g_0,\dots,t_kg_k) = (t_0gg_0,\dots,t_kgg_k).$$

Topologise EG in such a way as to obtain a CW-complex – if G is finite there is no problem, since the k-skeleton $\mathrm{E}G^k$ contains only finitely many cells, and inherits a unique topology from the disjoint union. Write $\mathrm{E}G = \bigcup_k \mathrm{E}G^k$. The proof in Chapter 1 that over $\mathbb{Z}$ the standard resolution is exact, that it has non-trivial homology only in dimension zero, carries over to show that E is contractible. Now define

$$\mathrm{B}G = \mathrm{E}G/G;$$

$\pi_1(\mathrm{B}G) \cong G$, since G-action permutes the cells and is hence free. The space BG is called the classifying space for the discrete group G, and the construction proves

Proposition 6.1

If A is a trivial G-module, then $H^k(\mathrm{B}G, A) \cong H^k(G, A)$.

The left hand side is to be interpreted as cellular cohomology and the right as the abstract cohomology group defined in Chapter 1.

Assume now that the order of G is finite.

Consider the monoid of positive representations $R^+(G)$ and the monoid of complex vector bundles $\mathrm{Vect}_*^{\mathbb{C}}(\mathrm{B}G)$, both admitting the operations $\oplus$ and $\otimes$. The flat bundle construction defines a map

$$\alpha: R^+(G) \to \mathrm{Vect}_*^{\mathbb{C}}(\mathrm{B}G)$$
$$\rho \longmapsto (\mathrm{E}G)\rho_\natural$$

which is compatible with sums, products and exterior powers. For any space X $\mathrm{Vect}_*^{\mathbb{C}}(X)$ maps into $K(X)$, the free abelian group of isomorphism classes of vector bundles modulo the subgroup generated by $\{(\xi_1 \oplus \xi_2) - \xi_1 - \xi_2\}$. The map α extends to a ring homomorphism from $R(G)$ into $K(\mathrm{B}G)$, also called the flat bundle homomorphism.

Let $\phi: G_1 \to G_2$ be a homomorphism of finite groups, inducing a continuous cellular map $\mathrm{B}\phi: \mathrm{B}G_1 \to \mathrm{B}G_2$ and hence a homomorphism of monoids

$$\mathrm{B}\phi^!: \mathrm{Vect}_*^{\mathbb{C}}(\mathrm{B}G_2) \to \mathrm{Vect}_*^{\mathbb{C}}(\mathrm{B}G_1).$$

The coordinate transformations for an induced bundle, and the map $\phi^!: R^+(G_2) \to R^+(G_1)$ are defined by composition with $\mathrm{B}\phi$ and ϕ respectively, so that the diagram below commutes:

$$\begin{array}{ccc} R(G_2) & \xrightarrow{\phi^!} & R(G_1) \\ \downarrow{\scriptstyle \alpha_2} & & \downarrow{\scriptstyle \alpha_1} \\ K(\mathrm{B}G_2) & \xrightarrow{\mathrm{B}\phi^!} & K(\mathrm{B}G_1) \end{array}$$

Next consider the bundle theoretic analogue of the transfer construction for representations, and suppose that $\mathrm{B}\phi$ is induced by a monomorphism. More generally one can take $h: X^1 \to X$ to be a finite covering map, but the special case is all that we will need. The transfer of a representation from the subgroup G_1 to G_2 can be described in the following classical way.

If F is a subgroup of the symmetric group S_s, $s = [G_2 : G_1]$, then the Wreath product $G_1 \wr F$ is defined as the semidirect product of $G_1 \times \cdots \times G_1$ (s factors) and F, where F acts by permuting the factors. The group G_2 may be embedded in $G_1 \wr S_s$ by first choosing a left transversal $\{1 = g_1, \ldots, g_s\}$, and noting that $y g_j = g_{\sigma(j)} x_j$ for each element of $y \in G_2$, some $x_j \in G_1$ and some permutation σ of the set $(1, 2, \ldots, s)$. Map

$$y \text{ to } (x_1, \ldots, x_s, \sigma)$$

in the semidirect product. From the definition of the induced representation module as the tensor product $\mathbb{C}G_2 \bigotimes_{G_1} M$, it is clear that if we first define an sn-dimensional module $(\rho \wr 1)(\mathbb{C}^n)$ for the group $G_1 \wr S_s$, and then restrict to the subgroup G_2, we obtain the module underlying $\phi_! \rho$. The notation $\rho \wr 1$ describes the action in which G_1 acts via ρ on each n-dimensional block of $\mathbb{C}^{ns} \cong \mathbb{C}^n \oplus \cdots \oplus \mathbb{C}^n$ and S_s permutes the blocks. There is a similar construction for vector bundles over classifying spaces, which we state in terms of G_1-vector bundles, see [Sg]. If p_{G_1}: $\mathrm{E}G_1 \to \mathrm{B}G_1$ is the projection map and ξ is an n-dimensional bundle over $\mathrm{B}G_1$, the $p_{G_1}^! \xi$ admits a G_1-action. Let S_s act on the contractable space

$$\underbrace{EG_1 \times \cdots \times EG_1}_{s} = (EG_1)^s$$

by permuting the factors, so that by combining the G_1 – and S_s – actions we obtain a free action by G_2 on $(EG_1)^s \times ES_s$. This extends to the total space of the product

$$p^!_{G_1}\xi \times \cdots \times p^!_{G_1}\xi,$$

and by passing to the space of orbits we obtain an ns-dimensional bundle over a homotopically equivalent model for the classifying space BG_2 constructed above. This defines the direct image homomorphism

$$\phi_!: \mathrm{Vect}^{\mathbb{C}}_*(BG_1) \to \mathrm{Vect}^{\mathbb{C}}_*(BG_2).$$

This construction formalises the intuitive idea of what the direct image should be – namely the fibre over each point of BG_2 consists of s copies of $\mathbb{C}^n$ indexed by the transversal $\{g_1, \ldots, g_s\}$. Furthermore $\phi_!$ induces a $K(BG_2)$-module homomorphism from $K(BG_1) \to K(BG_2)$, explicitly

$$\phi_!(\phi^!\xi \otimes \eta) = \xi \otimes \phi_!\eta.$$

To see this note that the pull-back of $\phi^!\xi$ over EG_1 is G_2-invariant. Thus

Proposition 6.2

The flat bundle homomorphism α is a natural transformation of Frobenius functors.

Definition

Let $\rho: G \to GL(n, \mathbb{C})$ be an n-dimensional representation of the finite group G, and $p_G: EG \to BG$ the universal principal G-bundle. The kth Chern class of ρ is the kth Chern class of the n-dimensional flat bundle $\alpha(\rho) = (EG)\rho_{\natural}$.

By Proposition 6.1 the kth Chern class takes values in $H^{2k}(G, \mathbb{Z})$ with trivial G-action on $\mathbb{Z}$. Transposing the properties stated in 5.10 to representation modules we have:

(1) $c_k(\rho) = 0$, $k > n$,
(2) if $\phi: G_1 \to G_2$ is a homomorphism $c_k(\phi^!\rho) = \phi^* c_k(\rho)$,
(3) if $c.(\rho) = 1 + c_1\rho + \cdots + c_n\rho$, $c.(\rho_1 \oplus \rho_2) = c.(\rho_1)c.(\rho_2)$, and
(4) $c_1: (\mathrm{Hom}(G, \mathbb{C}^\times), \otimes) \xrightarrow{\simeq} H^2(G, \mathbb{Z})$ is an isomorphism.

Only (4) perhaps needs further comment. Consider the exact coefficient sequence

$$\mathbb{Z} \rightarrowtail \mathbb{C} \underset{\exp(2\pi i\cdot)}{\twoheadrightarrow} \mathbb{C}^{\times}$$

and note that, since $\mathbb{C}$ is a divisible abelian group, $H^k(G,\mathbb{C})=0$ for $k\geqslant 1$. Hence the connecting homomorphism

$$\delta: H^1(G,\mathbb{C}^{\times}) = \mathrm{Hom}(G,\mathbb{C}^{\times}) \to H^2(G,\mathbb{Z})$$

is an isomorphism. This amounts to saying that the composition

$$\mathrm{Hom}(G,\mathbb{C}^{\times}) \underset{\alpha}{\to} \mathrm{Vect}_1^{\mathbb{C}}(BG) \underset{c_1}{\to} H^2(G,\mathbb{Z})$$

is an isomorphism.

At this point it is convenient to mention another property. In the first section of Chapter 5, we defined the exterior powers and Adams operations in $R(G)$. Given the definition of the exterior power of a vector bundle, one may formally introduce Adams operations ψ^j into the ring $K(X)$, in such a way that ψ^j is compatible with the homomorphism α.

(5) $c_k(\psi^j\rho) = j^k c_k(\rho)$.

Since $c_k(\rho)$ is the characteristic class of a bundle, we may work with bundles and apply the splitting principle (5.8). Therefore it suffices to consider the case $\xi = \eta_1 \oplus \cdots \oplus \eta_n$, a direct sum of line bundles.

From the definitions

$$\psi^j(\eta) = \eta^j \quad \text{and} \quad \psi^j(\xi) = \eta_1^j \oplus \cdots \oplus \eta_n^j.$$

Thus $c.(\psi^j\xi) = c.(\eta_1^j \oplus \cdots \oplus \eta_n^j)$

$$= (1 + jc_1\eta_1)\ldots(1 + jc_1\eta_n),$$

from which it is clear that the component in degree $2k$ equals

$$j^k c_k(\eta_1 \oplus \cdots \oplus \eta_n).$$

Property (5) now follows.

It would be extremely pleasing to have an axiomatic characterisation of the Chern classes of a representation, which would remove the necessity of using the topology of vector bundles over BG. One problem is the absence of a sufficiently strong splitting principle for *flat* bundles, that is one for which the splitting map $h: X^1 \to X$ induces a monomorphism for *integral* cohomology. With finite coefficients the situation is rather better, see the thesis of U. Staffelbach [St] and Chapter 9 below. However the non-abelian group P_2 of order p^3 and exponent p is such that the integral cohomology contains nilpotent elements which restrict to zero on every proper subgroup, see Chapter 8 below. This suggests that even where the map h exists, the class of spaces over which the flat bundles are defined is larger than the class $\{BK: K \subseteq G\}$.

Let us consider the problem of characterisation from a different direction. By Theorem 3.3 it is enough to consider groups G of prime power order, for which by Theorem 5.5 we know that an irreducible representation is either one-dimensional or the transfer of some one-dimensional representation. Since properties (1)–(4) determine the Chern classes of a sum of one dimensional representations, as in the outline proof of 5.11, in order to describe $c.(\rho)$ in general, we need a formula for the kth class of a transferred representation. In principle such a formula exists, see [Ev. 3]; it depends on induction over the index s and on knowing the classes of the permutation representation of S_r for $r \leqslant s$. (Chapter 7 is devoted to the discussion of these.) In practice Even's formula is quite unusable, but the argument suggests the following result.

Theorem 6.3
(Riemann–Roch formula for group representations)

Let $$M_k = \left(\prod_{p \leqslant k+1} p^{[k/p-1]} \right) \Big/ k!,$$

$$\bar{M}_k = \prod_{p|M_k} p.$$

and let s_k *be the kth Newton polynomial in the Chern classes,*

$$s_k(\rho) = N_k(c_1(\rho), \ldots, c_k(\rho)). \quad \text{Then}$$
$$\bar{M}_k(s_k(i_! \rho) - i_*(s_k \rho)) = 0.$$

Note that as usual $[a]$ means the integral part of a, and that no prime divides $\bar{M}_k$ with power greater than the first.

We postpone the proof until Appendix 1, since in its most elegant form this employs advanced homotopy theory, and even then restrict attention to the coarser multiple M_k. So far as the applications go, this involves little loss of generality, see in particular Theorem 8.6 and Proposition 7.2 below. The first, full, published proof of the theorem in its sharp, quoted form is the work of L. Evans & D. Kahn [E–K 2].

In the case of p-groups M_k may be replaced by a more elementary coefficient, since $H^*(G_p, \mathbb{Z})$ is p-torsion, and only powers of p dividing M_k are significant.

Proposition 6.4
If $k=(m_0+m_1p+m_2p^2+\cdots+m_rp^r)p^s$, $0\leqslant m_i\leqslant p-1$ and $m_0>0$, is the expansion of k to the base p, then the highest power of p to divide M_k is

$$\left[\frac{m_0+m_1+\cdots+m_r}{p-1}\right].$$

In particular, if $k=(p-1)p^s$, then M_k is divisible by p, but not by p^s. If $k=p^s$, then $(M_k,p)=1$.

Proof. The highest power of p to divide $k!$ equals

$$\left[\frac{k}{p}\right]+\left[\frac{k}{p^2}\right]+\cdots+\left[\frac{k}{p^{r-s}}\right].$$

In terms of the expansion to the base p with $l=m_0+m_1p+\cdots+m_rp^r$, the total power of p which occurs is therefore

$$\frac{l(p^s-1)}{(p-1)}+\left[\frac{l}{p}\right]+\cdots+\left[\frac{l}{p^r}\right].$$

But

$$\left[\frac{l}{p}\right]=m_1+m_2p+\cdots+m_rp^{r-1},$$

$$\left[\frac{l}{p^2}\right]=m_2+\cdots+m_rp^{r-2},$$

$$\left[\frac{l}{p^r}\right]=m_r.$$

Therefore the sum of the integral parts is

$$m_1\left(\frac{p-1}{p-1}\right)+m_2\left(\frac{p^2-1}{p-1}\right)+\cdots+m_r\left(\frac{p^r-1}{p-1}\right),$$

so adding $l(p^s-1)/(p-1)$ we get $(1/(p-1))(lp^s-(m_0+m_1+\cdots+m_r))$. Comparing this with $[lp^s/(p-1)]$ we see that the difference equals $[(m_0+m_1+\cdots+m_r)/(p-1)]$. The particular cases are now clear. Furthermore the argument shows that M_k is an integer rather than a rational number, and illustrates the extent to which $\bar{M}_k$ improves on M_k. The irregular pattern of the powers of p occurring is shown by Table 6.1 ($94\leqslant k\leqslant 125$, $p=5$), and is explained by the relative positions of the entries m_i between 0 and $p-1$.

Table 6.1

k	94	94	$96 \leqslant k \leqslant 99$	$100 \leqslant k \leqslant 103$	104	$105 \leqslant k \leqslant 107$	108, 109
Power of p dividing M_k	2	1	2	1	2	1	2

k	110, 111	$112 \leqslant k \leqslant 114$	115	$116 \leqslant k \leqslant 123$	124	125
Power of p dividing M_k	1	2	1	2	3	0

Definition

The Chern subring of the integral cohomology of G equals the subring of $H^{\text{even}}(G, \mathbb{Z})$ generated by the Chern classes of the irreducible complex representations of G. Denote this subring by $\text{Ch}(G)$.

Remark

If G is an elementary abelian p-group of rank greater than or equal to three, then $\text{Ch}(G)$ is properly contained in $H^{\text{even}}(G, \mathbb{Z})$.
In general $H^*(G, \mathbb{Z})$ is a finitely generated module over the Chern subring. Since G is finite the existence of the regular representation shows that there exists at least one faithful complex representation ϕ of G, and without loss of generality we may suppose that ϕ takes values in the compact subgroup $\text{U}(n)$. The method sketched in the final section of Chapter 5 shows how to construct the classifying space $\text{BU}(n)$, which is homotopy equivalent to $\text{BGL}(n, \mathbb{C})$. (The homogeneous space of positive definite Hermitian $n \times n$ matrices, $\text{GL}(n, \mathbb{C})/\text{U}(n)$, is contractable.) Hence $H^*(\text{BU}(n), \mathbb{Z})$ is polynomial in the universal Chern classes, and is Noetherian. The homomorphism $\text{B}\phi^*: H^*(\text{BU}(n), \mathbb{Z}) \to H^*(G, \mathbb{Z})$ (G is discrete :) gives $H^*(G, \mathbb{Z})$ the structure of a module over this Noetherian ring.

Theorem 6.5

$H^*(G, \mathbb{Z})$ is finitely generated over H^* $(\text{BU}(n), \mathbb{Z})$.

Proof. Once more this is a sketch only, see [Q2] for the details. Since G is a subgroup of $\text{U}(n)$ there is a fibration

$$\text{U}(n)/G \to BG \to \text{BU}(n),$$

in which the total space BG is homotopy equivalent to the model of BG constructed above. Consider the spectral sequence

$$E_2^{i,j} = H^i(\mathrm{BU}(n), H^j(\mathrm{U}(n)/G, \mathbb{Z})) \Rightarrow H^*(\mathrm{B}G, \mathbb{Z}).$$

Since $\mathrm{BU}(n)$ is 1-connected and its cohomology is a finitely generated free abelian group in each dimension,

$$E_2^{i,j} \cong H^i(\mathrm{BU}(n), \mathbb{Z}) \otimes H^j(\mathrm{U}(n)/G, \mathbb{Z}).$$

The rows $E_2^{*,j}$ possess an $H^*(\mathrm{BU}(n), \mathbb{Z})$-module structure, which is inherited by the terms of the filtration on page 39, and which is such that the differentials and the isomorphisms

$$E_{r+1} \cong H^*(E_r) \quad \text{and} \quad E_\infty^{*,j} \cong F^j H^*(\mathrm{B}G, \mathbb{Z})/F^{j+1} H^*(\mathrm{B}G, \mathbb{Z})$$

are compatible with the module structure.

Since $\mathrm{U}(n)/G$ is compact, $H^*(\mathrm{U}(n)/G, \mathbb{Z})$ is finitely generated, so that E_2^{**} is a Noetherian module. Therefore $E_r^{**}, 2 \leqslant r \leqslant r_0$, is also Noetherian and $E_{r_0}^{**} = E_\infty^{**}$. It follows that $H^*(\mathrm{B}G, \mathbb{Z})$ is a finitely generated module over the subring generated by the Chern classes of the faithful representation ϕ.

For the subring $\mathrm{Ch}(G) \subseteq H^{\mathrm{even}}(G, \mathbb{Z})$ one has the following numerical information:

Let G be a p-group of order p^n, containing a maximal normal abelian subgroup of order p^a, where the least value of a satisfies $a(a+1) \geqslant 2n$ [Hp, III, Satz 7.3]. If $a' = n - a$, then the general theory of representations implies that the maximum degree of an irreducible representation is $p^{a'}$, see the remark at the end of 5.3. Write p^e for the exponent of G.

Proposition 6.6

(i) $\mathrm{Ch}(G)$ *is generated in dimensions less than or equal to* $2p^{a'}$.

(ii) *The exponent of* $\mathrm{Ch}(G)$ *divides* $p^{a'+e+1}$.

Proof. Part (i) is immediate from the definitions. For (ii) note that p^e is *a fortiori* an exponent for the abelianisation of each subgroup of G. If $\hat{\gamma}$ is a 1-dimensional representation of G or of some proper subgroup, then $p^e i_*(\gamma^k)$ vanishes for $\gamma = c_1(\hat{\gamma})$. By the Riemann–Roch formula (Theorem 6.3 with the factor $\bar{M}_k$) and Blichfeldt's Theorem (5.5) p^{e+1} is an exponent for the subring of $\mathrm{Ch}(G)$ generated by the Newton polynomials s_j. The recurrence formula for the Newton polynomials now shows that if p^{e+1} annihilates the classes s_j, then $p^{a'+e+1}$ annihilates the underlying Chern classes.

Remarks

The bound in (ii) is weak, particularly as the simple argument following 2.6 shows that the exponent of an integral cohomology group divides p^n. However there are examples, of which the non-abelian group of order p^3 and exponent p is one, to show that $p^{a'+e}$ is the best value, see Appendix 2. The extra factor p is of course contributed by 6.3. It is also interesting to compare the exponent for Ch(G) with that obtained for $H^{\text{odd}}(G, \mathbb{Z})$ in [Ku]. If this latter exponent equals p^f, then p^{2f} divides the order of G.

Theorem 6.3 is a useful aid to calculation in low dimensions for small values of a', since

$$(M_k, p) = (\bar{M}_k, p) = 1 \quad \text{if} \quad k < p - 1.$$

It follows from 5.5 and elementary numerical considerations that, in dimensions less than $p-1$, Ch(G) is generated by the corestrictions of powers of 2-dimensional elements.

Before discussing some examples let us summarise the remaining widely used characteristic classes of representations. For an orientated real representation there are the Pontrjagin classes

$$p_k(\rho) = c_{2k}(EG\rho_\natural \otimes \mathbb{C}) \in H^{4k}(G, \mathbb{Z}) \quad \text{(stable)}$$

and the Euler class

$$e(\rho) = e(EG\rho_\natural) \in H^n(G, \mathbb{Z}) \quad \text{(unstable)}$$

For an arbitrary real representation $\rho: G \to \mathrm{GL}(n, \mathbb{R})$ the analogues of the Chern classes are the Stiefel–Whitney classes

$$w_k(\rho) = w_k(EG\rho_\natural) \in H^k(G, \mathbb{F}_2);$$

These are of special interest when G is a 2-group. As for bundles they are stable, that is unchanged by the addition of copies of the trivial representation, natural, and satisfy the exponential law $w.(\rho_1 \oplus \rho_2) = w.(\rho_1)w.(\rho_2)$.

First calculations

Restrict attention to groups of prime power order.

1. Let $\mathfrak{t}: C^T_{p^n} \to \mathbb{C}^\times$ be the one-dimensional representation of the cyclic group of order p^n, which maps T to a primitive p^n-th root of unity. The class $\tau = c_1(\mathfrak{t}) \in H^2(C_{p^n}, \mathbb{Z})$ generates the cohomology ring.

Let $D^*_{2^n}$ be the binary dihedral group introduced in the example–exercise at the end of Chapter 3. An irreducible representation of $D^*_{2^n}$ is either one-dimensional or the transfer of a representation of the index two subgroups generated by A. If ξ is a primitive 2^{n-1}st root of unity a typical such 2-dimensional representation $\hat{\xi}$ maps A to

$$\begin{pmatrix} \xi & 0 \\ 0 & \bar{\xi} \end{pmatrix}$$

and B to

$$\begin{pmatrix} 0 & 1 \\ -1 & 0 \end{pmatrix}.$$

Recall from the earlier calculation that $H^*(D^*_{2^n}, \mathbb{Z})$ is generated by $\alpha = c_1(\hat{\alpha})$, $\beta = c_1(\hat{\beta})$ and a 4-dimensional class ξ of order 2^n. Naturality and the sum formula show that $c_2(\hat{\xi})$ restricts to a generator of $H^4(C^A_{2^{n-1}}, \mathbb{Z})$, so we may take $\xi = c_2(\hat{\xi})$. This discussion exhausts the possibilities for groups of prime power order with periodic cohomology. In all cases

$$\mathrm{Ch}(G) = H^{\mathrm{even}}(G, \mathbb{Z}) = H^*(G, \mathbb{Z}).$$

2. Let G be the split metacyclic p-group of order p^{n+1} discussed in 4.4:

$$G = \{A, B: A^{p^n} = B^p = 1, A^B = A^{1+p^{n-1}}\}.$$

Let $i_!\hat{\alpha}$ denote an irreducible representation of degree p transferred up from a 1-dimensional representation $\hat{\alpha}$ mapping A to some primitive root of unity. In even dimensions smaller than $2(p-1)$

$$s_k(i_!\hat{\alpha}) = i_*(\alpha^k) = \alpha_k \text{ in the notation of 4.4.}$$

The group $H^{2p-2}(G, \mathbb{Z}) \cong \mathbb{Z}/p^{n-1} \times \mathbb{Z}/p$, where $i_*\alpha^{p-1}$ generates the first factor and β^{p-1} the second. In this dimension the Riemann–Roch formula gives

$$p(s_{p-1}(i_!\hat{\alpha}) - i_*(\alpha^{p-1})) = 0,$$

where the scalar p (see Lemma 8.7 below) is needed to kill a correction term lying in the β^{p-1}-factor. Hence, possibly by choosing a new generator, we may suppose that the factor of order p^{n-1} is generated by $s_{p-1}(i_!\hat{\alpha})$. In dimension $2p$, $(M_p, p) = 1$, so that

$$s_p(i_!\hat{\alpha}) = i_*(\alpha^p).$$

Using the recurrence formula for the Newton polynomials, together with the multiplicative relation

$$\alpha_i \alpha_j = p\alpha_{i+j}$$

from Theorem 4.4, we see that $pc_p(i_!\hat{\alpha})$ has order p^{n-1} in $H^{2p}(G, \mathbb{Z})$. Hence the top dimensional Chern class generates the summand in dimension $2p$ of order p^n. As in the previous example G is such that

$$\mathrm{Ch}(G) = H^{\text{even}}(G, \mathbb{Z}),$$

although there is now odd dimensional cohomology, which cannot be detected by Chern classes.

Exercise

Show that $H^{\text{even}}(D_{2^n}, \mathbb{Z})$ is also generated by Chern classes. Here D_{2^n} is the dihedral as opposed to the binary dihedral group already considered.

Our third example is of a group for which $\mathrm{Ch}(G)$ is properly contained in the even-dimensional cohomology, and for which the complete even dimensional structure has yet to be obtained. Since the example is important and our treatment of it fairly extended, we devote a new section to it.

Extra special *p*-groups

Recall that up to isomorphism there are two distinct non-abelian groups of order p^3, one metacyclic, the other (which we call elementary) of exponent p. An extra special p-group is obtained from the direct product of copies of either of these groups by amalgamating their centres, each of which is cyclic of order p. Up to isomorphism we may suppose that at most one 'factor' is metacyclic; however we shall confine our attention to the case where this is absent. Thus we assume that G has a presentation

$$G = \{A_i, B_i, C : [A_i, B_i] = C, A_i^p = B_i^p = C^p = [A_i, C] = [B_i, C] = [A_i, A_j] = [B_i, B_j] = [A_i, B_j] \quad (j \neq i) = 1\}.$$

The order of G equals p^{2n+1}; G is a central extension of a cyclic group of order p by an elementary abelian group of order p^{2n}, and contains a maximal elementary abelian subgroup K generated by $C, B_1, \ldots, B_n$. The irreducible representations have dimension 1 or p^n – a typical example of the latter type is $i_!(\hat{\gamma})$, where $\hat{\gamma}$ maps C to $e^{2\pi i/p}$ and B_i to 1 as a representation of K. It follows that the Chern ring is generated by 2-dimensional classes α_i, β_i corresponding to A_i, B_i and by the classes $\{c_k(i_!\hat{\gamma}) : 2 \leqslant k \leqslant p^n\}$.

Consider the commutative square

$$\begin{array}{ccc} H^{2k}(G,\mathbb{Z}) & \xrightarrow{j^*} & H^{2k}(G_i,\mathbb{Z}), \\ \uparrow{\scriptstyle i_*} & & \uparrow{\scriptstyle i_*|K_i} \\ H^{2k}(K,\mathbb{Z}) & \xrightarrow{j^*|K} & H^{2k}(K_i,\mathbb{Z}) \end{array}$$

in which $j: G_i \to G$ is the inclusion of the subgroup generated by $\{A_i, B_i, C\}$ of order p^3. When $k=1$, $\gamma = c_1(\hat{\gamma})$ restricts to an element γ_i, which is the first Chern class of the representation of K_i sending C to $e^{2\pi i/p}$ and B to 1. Either by direct calculation, see 8.6 below, or by referring to Appendix 2 on groups of order p^3, one shows that $i_*\gamma_i^k$ has order p for $k \geqslant 2$. Since $i_*\gamma^k$ has order at most p in $H^{2k}(G,\mathbb{Z})$ it follows that the order is precisely p. Furthermore the formula in Theorem 6.3 (see also Lemma 8.7 below) implies that

$$s_k(i_!\hat{\gamma}) = i_*(\gamma^k) + \text{correction terms.}$$

Since G has order divisible only by the prime p, $\bar{M}_k = 0$ or p, and in the former case all correction terms vanish, for example when $k < p-1$ or $k = p$. The multiplicative relations between the Newton polynomials may be read off from the corresponding relations in $H^*(G_i, \mathbb{Z})$, see Appendix 2 again, most of the products being zero. The same holds for products of the form

$$\alpha_i i_*(\gamma^k) \quad \text{and} \quad \beta_i i_*(\gamma^k).$$

Finally the recurrence formula for the Newton polynomials gives the rather surprising result that the additive order of $c_{p^m}(i_!\hat{\gamma})$ is p^{m+1}, $1 \leqslant m \leqslant n$. This value should be compared with the estimate for the exponent of the Chern ring given above in 6.6(ii).

Assume now that $p = 2$ and take coefficients in the field $\mathbb{F}_2$ with two elements, and consider the subring of $H^*(G, \mathbb{F}_2)$ generated by the Stiefel–Whitney classes of the real representations of G. Looking back at the description of the irreducible complex representations of G we see that, for $p = 2$, both the 1-dimensional representations $\hat{\alpha}_i$, $\hat{\beta}_i$ $(1 \leqslant i \leqslant n)$ and the 2^n-dimensional representation $i_!\hat{\gamma}$ are real.

Write G as the central extension

$$C_2 \rightarrowtail G \xrightarrow{\pi} V, \qquad V \cong \underbrace{F_2 \times \cdots \times F_2}_{2n}$$

which is classified by an element in $H^2(V, C_2)$, which in turn can be thought of as a quadratic function Q in the symmetric algebra $S(V^*)$ of the vector space dual to V over $\mathbb{F}_2$. Write B for the associated bilinear form. Then

$$iQ(\bar{x}) = x^2 \quad \text{if } \pi x = \bar{x}, \text{ and}$$
$$iB(\bar{x}, \bar{y}) = xyx^{-1}y^{-1} \quad \text{if } \pi x = \bar{x}, \pi y = \bar{y},$$

and $\pi^{-1}W$ is a (maximal) elementary abelian subgroup of G if and only if W is a (maximal) Q-isotropic subspace of V.

The mod 2 cohomology of G may be determined from the spectral sequence of the extension:

$$E_2^{i,j} \cong H^i(V, \mathbb{F}_2) \otimes H^j(C_2, \mathbb{F}_2) \Rightarrow H^*(G, \mathbb{F}_2).$$

By induction on $k \leqslant n$ it is possible to show that the elements

$$Q(\bar{x}), B(\bar{x}, \bar{x}^2), \ldots, B(\bar{x}, \bar{x}^{2^{k-1}})$$

in the cohomology of the base are hit from behind, and that if J_k equals the ideal which they generate, then

$$E_r^{**} \cong S(V^*)/J_k \otimes \mathbb{F}_2[w_{2^n}(i_!\hat{\gamma})], 2^{k-1}+1 < r \leqslant 2^k + 1.$$

In a manner similar to that seen in the calculation of the Chern ring above $w_{2^n}(i_!\hat{\gamma})$ restricts to a universal cycle in $H^{2^n}(C_2, \mathbb{F}_2)$. Hence $E_\infty = E_{2^n+1}$, and we have proved

Theorem 6.7 (D. Quillen)
$H^*(G, \mathbb{F}_2) \cong S(V^*)/J_n \otimes \mathbb{F}_2[w_{2^n}(i_!\hat{\gamma})]$. *In particular the mod 2 cohomology of an extra special 2-group is generated by Stiefel–Whitney classes.*

The last sentence follows because J_n is the kernel of the inflation map

$$\pi^*: H^*(V, \mathbb{F}_2) \to H^*(G, \mathbb{F}_2),$$

see Theorem 4.2, and the former ring is certainly generated by the first Stiefel–Whitney classes of a family of one dimensional real representations of the elementary abelian group V.

If $C_2 \times W = K$ is a maximal elementary abelian subgroup, one can show further that the non-zero Stiefel–Whitney classes of $i_!\hat{\gamma}|K$ are those in dimensions $2^n - 2^j$, for $0 \leqslant j \leqslant n$ and in dimension 2^n. For the details of all these calculations see [Q3].

Notes and references

The definition of the Chern classes of a representation as the

Chern classes of its associated flat bundle can be found in the Appendix to M.F. Atiyah's paper [At]. In this paper Atiyah observed that the missing component for an axiomatic characterisation was a good formula for c_k of an induced representation. In principle this is contained in [Ev3], and leads to a purely algebraic construction of the classes. However as a calculating device the formula for the Newton polynomials in 6.3 is much more fruitful. This formula has a somewhat chequered history – see the bibliography of the paper by L. Evens and D. Kahn [E-K2] for alternative versions.

The result on the finite generation of $H^*(G, \mathbb{Z})$ (6.5) is known as the Evens–Venkov Theorem, and has several published proofs. The one used here is due to D. Quillen – see [Q2]; the one due to L. Evens is purely algebraic and is related to his work on the Chern classes of an induced representation, [Ev1]. Both arguments use some of the properties of the multiplicative transfer in cohomology defined in [Ev2].

It is possible to apply the argument of 6.7 to extra special groups of odd prime power order; however one obtains only part of the mod p cohomology ring, see [Ya] and [Kr]. It is interesting to compare their calculations with the estimates of the size of Chern subring made above.

Problem

Determine the Chern subring of $H^{\text{even}}(G, \mathbb{Z})$ for the following two non-abelian groups of order P^4:

(a) $\{A, B \colon A^{p^2} = B^{p^2} = 1, A^B = A^{1+p}\}$ (Split metacyclic)

(b) $\{A, B, C \colon A^{p^2} = B^p = C^p = 1, [A, B] = C, [A, C] = [B, C] = 1\}$ (minimal non-abelian).

7

The symmetric group

The aim of this chapter is to determine the Chern subring of the integral cohomology of the symmetric group S_n.

Notation

Let S_n be the symmetric group on n symbols and

$$\pi_n\colon S_n \to \mathrm{GL}(n, \mathbb{C})$$

its representation by permutation matrices, i.e. $\pi_n(\sigma)$ equals the matrix obtained by permuting the rows of the unit matrix 1_n by means of σ. In order to determine the p-Sylow subgroup structure of S_n, write

$$n = r_m p^m + r_{m-1} p^{m-1} + \cdots + r_1 p + r_0.$$

Then the highest power of p which divides n, $v_p(n)$, equals

$$u = \left[\frac{n}{p}\right] + \left[\frac{n}{p^2}\right] + \cdots,$$

see page 70 above, and

$$u = r_m(p^{m-1} + p^{m-2} + \cdots + 1) \\ + r_{m-1}(p^{m-2} + p^{m-3} + \cdots + 1) + \cdots + r_1.$$

In particular $S_{p^m,p} = P_m$, a representative p-Sylow subgroup of S_{p^m} has order $p^{v(m)}$, where

$$v(m) = p^{m-1} + \cdots + 1,$$

and $S_{n,p}$ can be expressed as a direct product of groups of this type. Identify P_1 with the subgroup generated by the cycle $(1\,2 \ldots p)$; then P_2 is the Wreath product of the subgroup generated by the cycle $(1\,2 \ldots p)$

and the cyclic permutation group generated by $(1\,p+1\ldots p^2-p+1)$. Inductively

$$P_m = P_{m-1} \wr C_p,$$

see [Hp] or any basic book on group theory.

If $\pi_{p^m}^{(p)}$ denotes the restriction of π_{p^m} to the subgroup P_m, it is easy to see using the Wreath product construction that

$$\pi_{p^m}^{(p)} = \pi_{p^{m-1}}^{(p)} \wr 1 : P_m \to \mathrm{GL}(p^{m-1}, \mathbb{C}) \wr C_p \to \mathrm{GL}(p^m, \mathbb{C}).$$

As in the construction of the direct image bundle, page 66 above, inspection of the representing matrices shows that if $i = i_m : P_{m-1} \to P_m$ denotes the inclusion, then

$$\pi_{p^{m-1}}^{(p)} \wr 1 = i_!(\pi_{p^{m-1}}^{(p)}),$$

that is $\pi_{p^m}^{(p)}$ is obtained from $\pi_p^{(p)}$ by an m-fold application of the induced representation construction. When $m = 1$, $\pi_p^{(p)}$ equals the regular representation of the cyclic group $P_1 \cong C_p$.

Chern classes of the representation π_n

In what follows $\nu_p(k)$ equals the exponent of the prime p in the decomposition of k.

Theorem 7.1

If k is even and $n \gg k$ the highest power of p dividing the order of $c_k(\pi_n)$ equals

$$1 + \nu_p(k) \text{ if } k \equiv 0(p-1), \text{ and } 0 \text{ otherwise.}$$

Remark

Recall that one defines the Bernoulli numbers to be the coefficients in the power series

$$\frac{e^t}{e^t - 1} + \frac{t}{2} = 1 + \sum_{\substack{2 \leqslant k < \infty \\ k \text{ even}}} \frac{B_k}{k!} t^k.$$

Thus $B_2 = \frac{1}{6}$, $B_4 = -\frac{1}{30}$, $B_6 = \frac{1}{42}$, $B_8 = -\frac{1}{30}$, $B_{10} = \frac{5}{66}$, $B_{12} = -\frac{691}{2730}, \ldots$

By Von Staudt's theorem [B-Sh, Chapter 5, §8 Thm. 4], the highest power of p to divide the denominator of B_k/k is $1 + \nu_p(k)$ if $k \equiv 0(p-1)$. Hence Theorem 7.1 implies that the order of $c_k(\pi_n)$ divides $\mathrm{den}(B_k/k)$.

The proof consists of two parts. We first show the existence of an upper bound $\bar{E}_{\mathbb{Q}}(k)$ valid for any rational representation of any finite group, and then show that for the prime p this bound is attained by the representation π_n.

In the first section of Chapter 5, we claimed that if G is a finite group, and m a multiple of the exponent of G, then for an arbitrary automorphism σ of the complex numbers $\mathbb{C}$ there exists a positive integer j such that for all mth roots ω of 1

$$\sigma(\omega) = \omega^j.$$

In this case $\rho^\sigma = \psi^j\rho$ for all representations ρ of G, and if the character χ_ρ takes values in the subfield $\mathfrak{k}$ of $\mathbb{C}$, then for suitable values of j

$$\psi^j\chi_\rho = \chi_\rho.$$

This invariance with respect to the Adams operations ψ^j is the topological expression of invariance under the action of the Galois group $\mathrm{Gal}(\mathbb{C}/\mathfrak{k})$. Furthermore property (5) of the Chern classes of a representation reads

$$c_k(\rho^\sigma) = c_k(\psi^j\rho) = j^k c_k(\rho).$$

Hence for a suitably chosen family of integers j, dependent on the field of definition of ρ,

$$(1 - j^k)c_k(\rho) = 0.$$

Referring to the diagram of Galois groups on page 55 this holds whenever $j \in \mathrm{Im}(\lambda_{k,m})$ contained in the units of the ring $\mathbb{Z}/m$. We obtain a general invariance formula for representations defined over $\mathfrak{k}$, independent of the particular group G, by passing to an infinite cyclotomic extension $\mathfrak{k}_\infty$ of $\mathfrak{k}$, and completing the diagram on page 55 to

$$\begin{array}{ccc} \mathrm{Gal}(\mathbb{C}/\mathfrak{k}) & \xrightarrow{\ \mu\ } & \\ \downarrow & & (\hat{\mathbb{Z}})^* = \prod_p \mathbb{Z}_p^*, \\ \mathrm{Gal}(\mathfrak{k}_\infty/\mathfrak{k}) & \xrightarrow{\ \lambda\ } & \end{array}$$

where $\mathbb{Z}_p^*$ denotes the units in the p-adic integers. It follows that an upper bound $\bar{E}_{\mathfrak{k}}(k)$ for the order of $c_k(\rho)$ for all $\mathfrak{k}$-representations of all finite groups G is given by

$$\bar{E}_{\mathfrak{k}}(k) = \min_{j \in \mathrm{Image}\,\lambda_{\mathfrak{k}}} (1 - j^k).$$

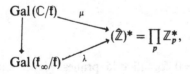

Now take $\mathfrak{k} = \mathbb{Q}$ and ρ to be the representation π_n of S_n, which is even defined over the integers. The lowest power of p needed to kill the p-torsion in $c_k\pi_n$ is given by the equation

$$v_p(E_{\mathbb{Q}}(k)) = \min_{j\in \mathbb{Z}_p^*}(\operatorname{ord}_p(1-j^k)).$$

There are two cases to consider. First let p be an odd prime. The group of multiplicative units decomposes as the direct sum

$$\mu_{p-1}\oplus(1+p\mathbb{Z}_p),$$

where μ_{p-1} denotes the cyclic subgroup of roots of unity. If $k\not\equiv 0 \pmod{(p-1)}$, $v_p(k)=0$. On the other hand if $p-1$ divides k, so that k is necessarily even,

$$v_p(k) = \min_{j\in 1+p\mathbb{Z}_p}(\operatorname{ord}_p(1-j^k)),$$

which on taking logarithms becomes

$$\begin{aligned} v_p(k) &= \min_{j\in p\mathbb{Z}_p}(\operatorname{ord}_p(kj)) \\ &= 1+\operatorname{ord}_p(k). \end{aligned}$$

When $p=2$ the argument is slightly different, since

$$\mathbb{Z}_2^* = \{\pm 1\}\oplus U,$$

where U is identified with $4\mathbb{Z}_2 \subseteq 2\mathbb{Z}_2$ by first squaring and then taking logarithms. The point is that the 2-adic exponential is only defined on $4\mathbb{Z}_2$; however log is still a monomorphism onto its image. Therefore if 2 divides k

$$\begin{aligned} v_2(k) &= \min_{j\in 1+4\mathbb{Z}_2}(\operatorname{ord}_2(1-j^k)) \\ &= \operatorname{ord}_2(4k) \\ &= 2+\operatorname{ord}_2 k. \end{aligned}$$

If k is odd $v_2(k)=1$ and $v_p(k)=0$ for all odd primes. Hence the odd dimensional Chern classes of π_n have order at most 2; this is a familiar topological property of real vector bundles.

We summarise the first part of the argument by:

$$p=\text{odd} \quad v_p(E_{\mathbb{Q}}(k)) = \begin{cases} 0 & k=\text{odd or } k\not\equiv 0\,(p-1), \\ 1+v_p(k) & k\equiv 0\,(p-1), \end{cases}$$

$$p=2 \quad v_2(E_{\mathbb{Q}}(k)) = \begin{cases} 1 & k=\text{odd}, \\ 2+v_2(k) & k=\text{even}. \end{cases}$$

In order to show that at least for odd primes this bound is attained by the representation π_n it is enough to consider p-torsion when k is divisible by $p-1$. With the notation introduced in the first section we have:

Lemma 7.2
For all primes p

$$v_p(\text{order of } c_{p^{m-1}(p-1)}\pi_{p^m}) = m = 1 + v_p(p^{m-1}(p-1)).$$

Proof. This is by induction on the exponent m. When $m=1$, $\pi_p^{(p)}$ equals the regular representation (with underlying module $\mathbb{C}C_p$) of the cyclic group P_1, and

$$c.(\pi_p^{(p)}) = 1 - \gamma^{p-1},$$

where γ is the first Chern class of the usual faithful representation of P_1. The result holds in this case; assume it for $1 \leqslant r \leqslant m-1$.

One interpretation of the formula proved in Theorem 6.3 is that

$$s_{p^{m-1}(p-1)}\pi_{p^m}^{(p)} = i_*(s_{p^{m-1}(p-1)}\pi_{p^{m-1}}^{(p)}) + \text{correction terms.}$$

Each of these correction terms has order at most p, and $\bar{M}_k$ is divisible by p for $k \equiv 0 \pmod{(p-1)}$ but not by p^2. Note in passing that the numerical considerations following the statement of 6.3 show that for $k = p^{m-1}(p-1)$ the scalar coefficient M_k is good enough. Hence the proof of the weaker form of 6.3 given in Appendix 1 applies here. Each non-zero correction term is detected by a term on the E_2^{**} page of the spectral sequence of a Wreath product extension (see 8.7 in the next chapter); in particular since the extension is split the non-zero term in

$$E_2^{2p^{m-1}(p-1),0} \cong H^{2p^{m-1}(p-1)}(P_1, \mathbb{Z})$$

survives to infinity. Hence at least one of the correction terms has order equal to p.

For all values of m and k the Newton polynomial $s_k(\pi_{p^{m-1}}^{(p)})$ has order at most k. This can be proved either by a subsidiary induction argument, using the recurrence formula

$$s_k - c_1 s_{k-1} + \cdots + (-1)^k k c_k = 0,$$

combined with the known upper bound for the order of c_k, or as follows:

$$s_k(\pi_{p^{m-1}}^{(p)}) = i_* s_k(\pi_{p^{m-2}}^{(p)}) + \text{correction terms of order at most } p.$$

Iterating this procedure and using the elementary fact that

$$i_{*\,K_2\to K_3}\, i_{*\,K_1\to K_2} = i_{*\,K_1\to K_3}$$

one sees that $s_k(\pi^{(p)}_{p^{m-1}})$ equals the corestriction of $s_k\pi^{(p)}_p$ up to correction terms of order p. Since the order of P_1 is p, the conclusion follows. We have given both arguments for this step, since the second uses the full force of 6.3 with $\bar{M}_k$ rather than with M_k, and this version must be used when $p=2$. Because of the extra factor 2 in the upper bound $\bar{E}_{\mathbb{Q}}(k)$ in this case the first argument shows that $s_k(\pi^{(2)}_{2^{m-1}})$ has order at most 4.

In dimension $p^{m-1}(p-1)$ the recurrence formula for the Newton polynomial reads

$$s_{p^{m-1}(p-1)} - c_1 s_{p^{m-1}(p-1)-1} + \cdots + p^{m-1}(p-1)c_{p^{m-1}(p-1)} = 0.$$

The restriction of the total Chern class $c.(\pi^{(p)}_{p^m})$ to a maximal p-torus shows that on such a torus the non-zero Chern classes are expressible as elementary symmetric functions in elements $t_i (1 \leqslant i \leqslant p^m)$ each of which has degree $2(p-1)$. Hence at least over $\mathbb{F}_p$ the classes $c_k(\pi^{(p)}_{p^m})$ are linearly independent for $k \equiv 0(p-1)$, and the upper bound already found shows that all others vanish. It follows that the relation above between the Newton Polynomials and the Chern classes is the most economical, that is, the order of the top-dimensional Chern class equals p times the p-primary factor of $p^{m-1}(p-1)$, as claimed.

If in the hypothesis of the theorem $k = p^{m-1}(p-1)$ and $n \geqslant p^m$ the conclusion now follows, since the maximal value of the p-primary part of $c_k(\pi_n)$ is certainly attained after restriction to the subgroup S_{p^m}. Recall that the class c_k is stable, that is, unchanged by the addition of copies of the trivial representation. For numbers of the form $k = lp^{m-1}(p-1)$ with $(l,p)=1$ consider the embedding

$$S_{p^m} \times S_{p^m} \times \cdots \times S_{p^m} \rightarrowtail S_{lp^m},$$

in which the first factor permutes the symbols $(1\,2 \ldots p^m)$, the second the symbols $(p^m+1 \ldots 2p^m)$ and so on. The permutation representation π_{lp^m} restricts to a representation which we may write as $\pi_{p^m} \times \cdots \times \pi_{p^m}$ on this subgroup, and the order of $c_{lp^{m-1}}(\pi_{p^m} \times \cdots \times \pi_{p^m})$ may be estimated using the Künneth formula. The Chern class contains a term of the form

$$c_{p^{m-1}(p-1)}(\pi_{p^m}) \otimes \cdots \otimes c_{p^{m-1}(p-1)}(\pi_{p^m}),$$

which by 7.2 has order p^m at the prime p. Since $\bar{E}_{\mathbb{Q}}(lp^{m-1}(p-1))_{(p)} = p^m$, and we have shown that this power is achieved after restriction to a subgroup, the proof of the theorem is complete.

Remark

In [N] Nakaoka proves that the cohomology groups of the symmetric groups are stable with arbitrary trivial coefficients, that is

$$i^*: H^k(S_n, \mathbb{Z}) \to H^k(S_m, \mathbb{Z})$$

is an isomorphism for $n > m \geq 2k$. It follows that for large values of n the cohomology groups H^k are independent of the particular value chosen, something which we have explicitly proved for the classes $c_k(\pi)$ in 7.1. Indeed if S_∞ is defined as the group of finite permutations on the set of natural numbers;

$$\pi\colon S_\infty \to \mathrm{GL}(\mathbb{C})$$

is a representation of S_∞ in $\mathrm{GL}(\mathbb{C}) = \bigcup_n \mathrm{GL}(n, \mathbb{C})$, whose Chern classes in $H^{\text{even}}(S_\infty, \mathbb{Z})$ have the values stated in 7.1.

Having determined the orders of the Chern classes of π_n it is easy to deduce the structure of the Chern ring $Ch(S_n)$, using the following result:

Theorem 7.3

The complex representation ring $R(S_n)$ is generated as a λ-ring by the class of the natural representation π_n.

Proof. See [Bo].

Since the Chern classes of the exterior powers of a bundle or representation can be expressed in terms of the Chern classes of the original bundle, we have

Corollary

$\mathrm{Ch}(S_n)$ *is generated by the classes* $\{c_k\pi_n : k \leq n\}$ *and hence has exponent dividing the lowest common multiple of the numbers* $\{\mathrm{den}(B_k/k) : k \leq n \text{ and } k \text{ even}\}$.

Linear groups over rings of algebraic integers

As before suppose that $\mathfrak{k}$ is a finite extension field of the rational numbers $\mathbb{Q}$ and that $\mathfrak{k}_m$ is the mth cyclotomic extension of $\mathfrak{k}$, that is $\mathfrak{k}_m = \mathfrak{k}(\zeta)$ for ζ equal to some primitive mth root of unity. Write

$$\gamma_{\mathfrak{k}}(2) = \sup\{a : \mathfrak{k}_4 = \mathfrak{k}_{2^{a+1}}\},$$
$$\gamma_{\mathfrak{k}}(p) = \sup\{a : \mathfrak{k}_p = \mathfrak{k}_{p^a}\} \quad (p = \text{odd}), \text{ and}$$

$$\phi_{\mathfrak{k}}(m) = \text{degree of } \mathfrak{k}_m \text{ over } \mathfrak{k}.$$

For example $\phi_{\mathbb{Q}}(m) = \phi(n)$, the usual Euler function.

Definition

The field $\mathfrak{k}$ is *exceptional* if and only if $\mathrm{Gal}(\mathfrak{k}_{2^b}/\mathfrak{k})$ is cyclic for all exponents b. The distinction between exceptional and other fields is necessary because the group of units in the ring $\mathbb{Z}/2^b$ is isomorphic to $(\mathbb{Z}/2) \times (\mathbb{Z}/2^{b-2})$ rather than cyclic.

Examples

1. If $\sqrt{-1} \in \mathfrak{k}$, $\mathfrak{k}$ is exceptional.
2. If $\sqrt{-1} \notin \mathfrak{k}$, $\mathfrak{k}$ is exceptional if and only if the intersection $\mathfrak{k} \cap \mathbb{Q}_{2^{c+2}}$ is not contained in the real subfield $\mathbb{Q}(\cos(\pi/2^c))$.
3. The field $\mathfrak{k}$ is *formally real* if -1 is not a sum of squares in $\mathfrak{k}$. Formally real fields are not exceptional.
4. The only exceptional quadratic fields are $\mathbb{Q}(\sqrt{-1})$ and $\mathbb{Q}(\sqrt{-2})$. As in the first section of this chapter it is possible to define an upper bound $\bar{E}_{\mathfrak{k}}(k)$ for the order of the class $c_k(\rho)$, when ρ is defined over $\mathfrak{k}$. The general theory goes through unchanged, and subject to suitable identifications

$$\bar{E}_k(\mathfrak{k}) = \min_{j \in \mathrm{Gal}(\mathbb{C}/\mathfrak{k})} (1 - j^k).$$

One determines the exponent $v_p(\bar{E}_{\mathfrak{k}}(k))$ by allowing j to run through the elements of a certain subgroup of the p-adic units $\mathbb{Z}_p^*$, from which it is clear that $\bar{E}_{\mathbb{Q}}(k)$ divides $\bar{E}_{\mathfrak{k}}(k)$. Indeed the p-primary parts agree if p does not divide the discriminant of $\mathfrak{k}$. By a careful analysis of the Galois groups concerned B. Eckmann and G. Mislin obtain the following table [E-M] for the exponent of the p-primary part of $\bar{E}_{\mathfrak{k}}(k)$:

$$p = \text{odd} \begin{cases} k \not\equiv 0 \pmod{\phi_{\mathfrak{k}}(p)}, & v_p(\bar{E}_{\mathfrak{k}}(k)) = 0, \\ k \equiv 0 \pmod{\phi_{\mathfrak{k}}(p)}, & v_p(\bar{E}_{\mathfrak{k}}(k)) = \gamma_{\mathfrak{k}}(p) + v_p(k), \end{cases}$$

$$p = 2 \begin{cases} k = \text{odd} \begin{cases} v_2(\bar{E}_{\mathfrak{k}}(k)) = 1 \text{ if } \sqrt{-1} \notin \mathfrak{k}, \\ v_2(\bar{E}_{\mathfrak{k}}(k)) = \gamma_{\mathfrak{k}}(2) + 1 \text{ if } \sqrt{-1} \in \mathfrak{k}, \end{cases} \\ k = \text{even} \begin{cases} v_2(\bar{E}_{\mathfrak{k}}(k)) = \gamma_{\mathfrak{k}}(2) + \gamma_2(k) + 1, \\ \qquad \text{if } \sqrt{-1} \in \mathfrak{k} \text{ or if } \mathfrak{k} \text{ is not exceptional}, \\ v_2(\bar{E}_{\mathfrak{k}}(k)) = \gamma_{\mathfrak{k}}(2) + \gamma_2(k), \\ \qquad \text{if } \sqrt{-1} \notin \mathfrak{k} \text{ and } \mathfrak{k} \text{ is exceptional}. \end{cases} \end{cases}$$

As over the rationals $\bar{E}_{\mathfrak{k}}(k)$ differs from the best possible value $E_{\mathfrak{k}}(k)$ by at most a factor of 2, which occurs if k is even and $\mathfrak{k}$ is formally real. One proves this by constructing $\mathfrak{k}$ representations for suitable finite groups (cyclic groups of prime power order, (semi)dihedral and binary dihedral 2-groups suffice) for which the bound is attained.

As an illustrative example consider the case of quadratic fields. If $\mathfrak{k} \cap \mathbb{Q}_{p^n} = \mathbb{Q}$ for all prime powers p^n, then $\phi_{\mathfrak{k}}(p) = \phi_{\mathbb{Q}}(p)$ and $\gamma_{\mathfrak{k}}(p) = \gamma_{\mathbb{Q}}(p)$. The only quadratic number fields which occur as subfields of cyclotomic fields are

$$\mathbb{Q}(\sqrt{d}) \text{ for } d = \pm 1, \pm 2 \text{ and } (-1)^{(q-1)/2} q \text{ for } q = \text{odd prime}.$$

With these fields excluded

$$E_{\mathfrak{k}}(k) = \begin{cases} E_{\mathbb{Q}}(k), & k \text{ odd or } \mathfrak{k} \text{ real}, \\ 2E_{\mathbb{Q}}(k) & k \text{ even and } \mathfrak{k} \text{ imaginary}. \end{cases}$$

Among the excluded fields the Gaussian numbers $\mathbb{Q}(\mathrm{i})$ are such that

$$E_{\mathbb{Q}(\mathrm{i})}(k) = \begin{cases} 4, & k \text{ odd} \\ 2E_{\mathbb{Q}}(k), & k \text{ even}. \end{cases}$$

The last entry is illustrated by the natural representation of the group D_8^* in $GL(2, \mathbb{Z}(\mathrm{i}))$. As our earlier calculations have shown this has a second Chern class of order 8 rather than 4, see [Th5].

The results on Galois invariance extend to finite dimensional representations of infinite discrete groups over the complex numbers $\mathbb{C}$. However in order to do this one needs to put the theory of characteristic classes in a more general framework than the one developed in Chapter 6. In order to justify the use of the upper bound $\bar{E}_{\mathfrak{k}}(k)$ for representations of certain arithmetic groups, we must at least outline the more general theory, which is developed at length in [G] and [J].

Let $\rho: G \to \mathrm{Aut}(\mathbb{C}^n)$ be an n-dimensional complex representation of the discrete group G, possibly infinite. Independently of bases

$$(\det)(\rho) \in \mathrm{Hom}(G, \mathbb{C}^\times) \cong H^1(G, \mathbb{C}^\times).$$

If β is the coboundary map defined by the short exact sequence

$$\mu_{p^n} \rightarrowtail \mathbb{C}^\times \twoheadrightarrow \mathbb{C}^\times, \quad z \longmapsto z^{p^n}$$

let

$$c_1(\rho) = \beta((\det)\rho) \in H^2(G, \mu_{p^n}).$$

As n varies the classes obtained are compatible, and in the limit one may define the first p-adic Chern class $c_1(\rho)_p \in H^2(G_1 \mathbb{T}_p)$, where

$$\mathbb{T}_p = \varprojlim_n \boldsymbol{\mu}_{p^n} \cong \mathbb{Z}_p.$$

The last isomorphism is not unique. If $\mathscr{T}_G$ denotes the Grothendieck topology of finite G-sets and finite surjective coverings, the cohomology groups of $\{1\}$ with coefficients in the trivial sheaf $\mathbb{T}_p$ are isomorphic to the cohomology groups of G, and there is a (1–1) correspondence between $\mathbb{C}[G]$-modules and sheaves of vector spaces over $\mathscr{T}_G$. In this setting mimic the construction of the topological Chern classes by first defining them for sums of one-dimensional representations, and then using a splitting principle. (Compare the thesis of U. Staffelbach [St], where the main theorem is stated for flat rather than arbitrary vector bundles). In this way one obtains inside an algebraic framework unique classes

$$c_k(\rho)_p \in H^{2k}(G, \mathbb{T}_p^{\otimes k}),$$

satisfying the same axioms as before. In addition one has

(A) Comparison. If $c_k(\rho)^{\text{top}}_{(p^n)}$ denotes the mod p^n reduction of the class defined in Chapter VI, and one chooses compatible isomorphisms $\mathbb{Z}/_{p^n\mathbb{Z}} \cong \boldsymbol{\mu}_{p^n}$, then

$$c_k(\rho)_p = \varprojlim_n \left(c_k(\rho)^{\text{top}}_{(p^n)}\right),$$

that is, $c_k(\rho)_p$ carries the p-torsion of the topologically defined class.

(B) Galois invariance. If ρ is defined over the algebraic number field $\mathfrak{k}$, then the Galois action on the roots of unity induces an action on $\mathbb{T}_p^{\otimes k}$, with respect to which the algebraic classes are Gal$(\mathbb{C}/\mathfrak{k})$-invariant. This leads to exactly the same formula as before, namely

$$v_p(\bar{E}_{\mathfrak{k}}(k)) = \min_{j \in \mathrm{Gal}(\mathbb{C}/\mathfrak{k})} (\mathrm{ord}(1 - j^k)),$$

see [G] pages 257–8, but which is now valid, thanks to (A), for an arbitrary discrete group.

Let $\mathfrak{o}$ be the ring of integers in $\mathfrak{k}$ and

$$\sigma_n: \mathrm{GL}(n, \mathfrak{o}) \rightarrowtail \mathrm{GL}(n, \mathbb{C})$$

the inclusion representation. As with the symmetric group S_n the cohomology of $\mathrm{GL}(n, \mathfrak{o})$ is stable; so that for $n \gg k$, $H^{2k}(\mathrm{GL}(n, \mathfrak{o}), \mathbb{Z})$ is independent of n, see [Ch]. In the same range

$$\bar{E}_{\mathfrak{k}}(k) c_k(\sigma.) = 0,$$

and by restriction to finite subgroups one sees that this bound is always attained at odd primes, and also at the prime 2 if k is odd or $\mathfrak{k}$ not formally real. As a special case let $\mathfrak{k} = \mathbb{Q}$ and apply Theorem 7.1; then if k is even and $\sigma = \mathbb{Z}$, $c_k(\sigma.)$ has order equal to either den (B_k/k) or den$(B_k/2k)$. At odd primes it is enough to restrict to the representation π of the symmetric group; for the prime 2 the extra factor 2 is certainly needed for odd values of k, see [Ar]. Hence at this point the theory for infinite discrete groups begins to diverge from that for finite.

The following topological digression may help to put the last paragraph in perspective. If $K_{4k-1}(\mathbb{Z})$ is identified with the homotopy group $\pi_{4k-1}(B\mathrm{GL}(\mathbb{Z})^+)$, where $\mathrm{GL}(\mathbb{Z}) = \bigcup_n \mathrm{GL}(n, \mathbb{Z})$ and the plus construction gives a space with the same (co)homology as $\mathrm{GL}(\mathbb{Z})$, then the image of the J-homomorphism $J(\pi_{4k-1}0)$ in the $(4k-1)$st stable homotopy group of the sphere is isomorphic to a subgroup of $K_{4k-1}(\mathbb{Z})$, see [Q5]. If k is even this subgroup splits off as a direct summand; if k is odd the same holds for the odd torsion. Since $K_3(\mathbb{Z})$ is cyclic of order 48 and $J(\pi_3 0)$ has order equal to 24, the same cannot hold for the prime 2. Incidentally it is this calculation which shows that $c_2(\sigma^{\mathbb{Z}}_.)$ has order 24 rather than 12 [Ar].

The Chern class $c_k(\sigma_n) \in H^{2k}(\mathrm{GL}(n, \mathbb{Z}), \mathbb{Z})$ is an example of a torsion class, whose existence can be detected by the family of finite subgroups of $\mathrm{GL}(n, \mathbb{Z})$. However here is an explicit example to show that this need not be so, [So]. Consider the numerical function $M(n) = \prod_p p^{r(p,n)}$, where

$$r(p,n) = \begin{cases} \left[\dfrac{n}{p-1}\right] + \left[\dfrac{n}{p(p-1)}\right] + \left[\dfrac{n}{p^2(p-1)}\right] + \cdots, & p, \text{ odd} \\ n + \left[\dfrac{n}{2}\right] + \left[\dfrac{n}{4}\right] + \cdots, & p = 2. \end{cases}$$

Then $M(n)$ equals the lowest common multiple of the orders of finite subgroups of $\mathrm{GL}(n, \mathbb{Z})$, for example [NB, Exercises III, §7, 5–8]. Recent results in number theory imply that the numerator of the Bernoulli number B_k divides the order of $K_{4k-2}(\mathbb{Z})$; in particular $K_{22}(\mathbb{Z})$ contains an element of order 691 (see the first few values in the statement of Theorem 7.1 above). This element survives into $H_{22}(\mathrm{GL}(\mathbb{Z}, \mathbb{Z})$ under the Hurewicz map, but no dual cohomology class can be detected by a finite subgroup, since the stability theorem shows that it would suffice to take $n = 50$, and the numerical function $M(n)$ shows that $\mathrm{GL}(50, \mathbb{Z})$ contains no 691-torsion.

Notes and references

The proof of Theorem 7.1 is a collation from several sources. The use of the Adams operations to express Galois invariance is to be found in [A-T] and is systematically exploited in [E-M]. The calculation of $\bar{E}_{\mathbb{Q}}(k)$ is taken from [Th3]. Lemma 7.2 is a variant of the argument used in [E-K1]; it is comparatively simple because we can use both the existence of the upper bound $\bar{E}_{\mathbb{Q}}(k)$ to restrict attention to values of k satisfying $k \equiv 0(\mathrm{mod}\,(p-1))$, and the full force of the Riemann–Roch formula in Theorem 6.3 to estimate the orders of various correction terms. This hides the extensive use of spectral sequences made by L. Evens & D. Kahn. The survey in the second part of the chapter is intended to do no more than summarise what is known to me at the time of writing, and to illustrate the sort of information which can be obtained about the torsion in the cohomology of arithmetic groups by simple methods. In order to appreciate the full power of the algebraic definition of Chern classes for representations over a not necessarily closed field the interested reader should look at the work of C. Soulé, for example in [So].

8

Finite groups with p-rank $\leqslant 2$

At the end of Chapter 6 we gave a number of examples to show that it is possible for the subring of Chern classes to exhaust the even dimensional cohomology of a finite group. We propose to examine this phenomenon more carefully; as the examples suggest the relative 'size' of $\mathrm{Ch}(G)$ is closely connected to the abelian subgroup structure of G.

Definition

The p-rank (rk_p) of the finite group G equals the dimension over $\mathbb{F}_p$ of a maximal elementary abelian p-subgroup K of G. Thus if $K \cong (\mathbb{Z}/p) \times \cdots \times (\mathbb{Z}/p)$ (r copies), $rk_p(G) = r$.

If $rk_p(G) = 1$, a p-Sylow subgroup is either cyclic or generalised quaternion/binary dihedral (if $p = 2$). Thus $rk_p(G) = 1$ if and only if $H^*(G, \mathbb{Z})_{(p)}$ is periodic, see the final section of Chapter 3.

Abelian groups

Theorem 8.1

(i) $H^*(C_p, \mathbb{Z}) = (\mathbb{Z}/p)[\alpha]$, $\dim(\alpha) = 2$.

(ii) $H^*(C_p \times C_p, \mathbb{Z}) = (\mathbb{Z}/p)[\alpha, \beta] \otimes E(\mu)$, *if p is an odd prime,* $\dim(\alpha) = \dim(\beta) = 2$, *and* $\dim(\mu) = 3$.

(iii) $H^{\mathrm{even}}(C_p \times C_p \times C_p, \mathbb{Z})$ *is generated over $\mathbb{Z}/p$ by elements α, β and γ of dimension 2 and by ξ of dimension 4.*

Proof. Part (i) is familiar; parts (ii) and (iii) follow from the diagram

$$\begin{array}{ccccc} H^k(G, \mathbb{Z}) & \rightarrowtail & H^k(G, \mathbb{F}_p) & \twoheadrightarrow & H^{k+1}(G, \mathbb{Z}) \\ & & \Delta \downarrow & & \downarrow \\ & & & & H^{k+1}(G, \mathbb{F}_p) \end{array}$$

in which the two monomorphisms are reduction mod p. The long exact sequence of cohomology groups associated with the coefficient sequence $\mathbb{Z} \rightarrowtail \mathbb{Z} \twoheadrightarrow \mathbb{F}_p$ breaks up into short exact sequences, and $H^*(G, \mathbb{Z})$ coincides with the kernel of the derivation Δ. In particular in part (iii) the generator ξ arises from an $\mathbb{F}_p$-element of the form

$$x_1x_2y_3 - x_1y_2x_3 + y_1x_2x_3, \text{ where}$$
$$\Delta x_i = y_i \text{ and } \Delta y_i = 0, \text{ with due attention paid to signs.}$$

Since for any abelian group the irreducible complex representations are all 1-dimensional, the generators of the Chern ring all have degree 2. In case (iii) above ξ is independent of α, β and γ, so ξ is outside $\mathrm{Ch}(C_p \times C_p \times C_p)$. This example suggests the restriction to $rk_pG \leqslant 2$ in the sections which follow.

Groups with *p*-periodic cohomology

For the sake of completeness we start with

Lemma 8.2

If G has p-periodic cohomology $H^{\mathrm{odd}}(G, \mathbb{Z})_{(p)} = 0$.

Proof. By Theorem 3.3 the p-torsion subgroup is detected by $H^{\mathrm{odd}}(G_p, \mathbb{Z})$, which vanishes for both cyclic and binary dihedral groups.

Theorem 8.3

If $rk_p(G) = 1$, $H^*(G, \mathbb{Z})_{(p)} = H^{\mathrm{even}}(G, \mathbb{Z})_{(p)} = \mathrm{Ch}(G)_{(p)}$.

Proof. We have already shown in Chapters 3 and 7 that the theorem holds for groups of prime power order. In the case of C_{p^n} we may take $\alpha = c_1(\hat{\alpha})$ as a polynomial generator; in the case of $D^*_{2^n}$, $\alpha = c_1(\hat{\alpha})$, $\beta = c_1(\hat{\beta})$ and $\xi = c_2(\hat{\xi})$, where $\hat{\xi}$ is a faithful 2-dimensional representation, see page 35 above.

Suppose next that G has composite order and that p is an odd prime. A representative p-Sylow subgroup G_p is cyclic of order p^n, and by the normaliser/centraliser condition of Lemma 3.4

$$H^{\mathrm{even}}(G, \mathbb{Z})_{(p)} \cong H^{\mathrm{even}}(G_p, \mathbb{Z})^{N_p}.$$

Furthermore by Lemma 3.11 $d = d_p = 2[N_p{:}Z_p]$; by identifying G_p with $C^A_{p^n}$, and by setting $\hat{\alpha}(A) = e^{2\pi i/p^n}$, $\alpha = c_1(\hat{\alpha})$, we see that

$$H^{\mathrm{even}}(G, \mathbb{Z})_{(p)} \cong (\mathbb{Z}/p^n)[\alpha^{d/2}].$$

The problem now is to find an element of the representation ring $R(G)$ (an integral combination of representations, possibly with some negative multiplicities), the restriction of which to G_p has $(d/2)$-th Chern class equal to $\alpha^{d/2}$. Note that the argument in 3.4 is purely group theoretic, and therefore applies not only to more general cohomology theories than $H^*(\cdot, \mathbb{Z})$, but also to representations. Thus, if in the formulation of Brauer's Theorem on page 52 the elementary subgroup G_p is assumed to be abelian, then the image of $R(G)$ under restriction equals $R(G_p)^{N_p}$. The superscript N_p once more specifies those elements of the representation ring which are invariant under automorphisms of G_p induced by N_p-conjugation.

The representation $\hat{\alpha}$ of the cyclic group G_p extends to a one-dimensional representation ξ of the centraliser Z_p. Define

$$\eta = i_{N_p \to G_p} i_{!Z_p \to N_p}(\xi),$$

that is, first transfer ξ up to the normaliser and then restrict this representation to the Sylow subgroup G_p. The eigenvalues of the matrix representing A are equal to $e^{2\pi i q_j/p^n}$, $j = 1, 2, \ldots [N_p : Z_p]$, where

$$A^{X_j} = A^{q_j}$$

for some transversal $\{X_1 = 1, X_2, \ldots, X_{[N_p:Z_p]}\}$ of Z_p in N_p. Since the induced action on $R(G_p)$ of an element of N_p permutes these eigenvalues the representation η is N_p-invariant, that is, η belongs to the image of $i_{G \to G_p}$. The top-dimensional Chern class

$$c_{[N_p:Z_p]}(\eta) = q_1 q_2 \ldots q_{[N_p:Z_p]} \alpha^{d/2},$$

which generates $H^d(G_p, \mathbb{Z})$, since each multiple q_j is coprime with p^n.

For $p = 2$ there are two subcases to consider. Suppose first that G_2 is cyclic of order 2^n. Since G_2 is contained in Z_2, the index $[N_2 : Z_2]$ is odd, as is the order of the image of the subgroup of automorphisms in $\mathrm{Aut}(G_2)$ induced by N_2-conjugation. Since $\mathrm{Aut}(G_2)$ has order equal to a power of 2, the action of N_2 is trivial, and $H^*(G, \mathbb{Z})_{(2)}$ is generated in dimension 2 by a first Chern class, see page 67 property (4).

If G_2 is binary dihedral one must work a bit harder. First one needs a lemma similar to 3.4, due to R.G. Swan [Sw], and already used implicitly in the extended example on pages 34–5.

Lemma 8.4

If G_2 is a binary dihedral 2-Sylow subgroup of the finite group G, then $H^4(G_2, \mathbb{Z})$ is stable.

Proof. First note (using the particular resolution on page 35) that

$H^4(G_2, \mathbb{Z})$ is fixed under automorphisms of the form

$$A \longmapsto A, B \longmapsto A^i B.$$

If G_2 has order 8, every automorphism is of this type, and there is nothing more to prove. Inspection also shows that all embeddings of C_2, C_4 or D_8^* into G_2 induce the same map on H^4 – if $n \geqslant 4$, G_2 contains exactly two subgroups isomorphic to D_8^*. Suppose that G_2 has order at least 16, and write $G_{2,8}$ for the characteristic subgroup generated by A, containing all elements of order at least 8. Let T be the subgroup of G consisting of all elements U such that

$$A^U = A \text{ or } A^{-1};$$

then T contains both G_2 and G_2^X as subgroups. Since G_2 is maximal in both G and T, $G_2^{XU} = G_2$ for some $U \in T$, and replacing U by BU if necessary, we may suppose that $A^U = A$. We have so far shown that restriction following conjugation by X can be decomposed into the composition of conjugation by UX and the twisted restriction given by the embedding $g(Y) = Y^U$. Since U centralises the generator A, g is either the inclusion of the subgroup or the composition of the inclusion with one of the preferred automorphisms of G_2 already considered. It remains to show that conjugation by UX is trivial – if $UX \in G_2$ there is nothing to prove; otherwise conjugation by UX has odd order and is trivial in effect on $H^4(G_2, \mathbb{Z})$, a cyclic group of order a power of 2.

As in the case of groups of odd prime power order it is possible to mimic the steps of Lemma 8.4 on the set of irreducible 2-dimensional representations $\{\hat{\xi}_j : j = 1, 2, \ldots, 2^{n-2} - 1\}$ of $D_{2^n}^*$. If $t = 3$ there is only one such. By inspection it is even clearer than in the case of $H^4(G_2, \mathbb{Z})$ that a preferred automorphism of G_2 fixes the representation $\hat{\xi}_j$, and that any embedding of D_8^* into $D_{2^n}^*$ fixes $\hat{\xi} = \hat{\xi}_1$. The same holds for the image under restriction of the set of $\hat{\xi}_j$s for any embedding of C_2 or of C_4 into G_2. As above, when $n \geqslant 4$, one is reduced to considering the effect on the set of $\hat{\xi}_j$s of conjugation by $V = UX$, an element of the normaliser N_2. If $V \in G_2$ there is nothing to prove, hence the image of N_2 in the set of permutations of the $\hat{\xi}_j$ has odd order.

By looking at the effect of the image of N_2 on the set of restrictions of the representations $\hat{\xi}_j$ to the subgroup of index 2 generated by A, we see once more that the 2-primary order of $\operatorname{Aut}(\mathbb{Z}/2^{n-1})$ forces N_2 to act trivially.

The previous paragraph shows that for some representation $\hat{\xi}_j$ of G_2 there is at least one (virtual) representation of G which restricts to $\hat{\xi}_j$, and whose second Chern class therefore generates $H^*(G,\mathbb{Z})_{(2)}$. In dimensions congruent to 2 modulo 4 any torsion arising can be handled using the class c_1 as above. This concludes the proof of 8.3 in all cases.

Remark

If G is p-solvable and has p-periodic cohomology, S. Jackowski and T. Zukowski [J-Z] construct an explicit homomorphism of G into $GL(d_p/2,\mathbb{C})$, the top dimensional Chern class of which generates $H^{d_p}(G,\mathbb{Z})_{(p)}$. However p-solvability is essential to their argument. The alternating group A_5 is not 3-solvable, but it is well-known that there are two irreducible 3-dimensional representations, $\rho_{\pm}$, for which the representing matrix of a generator of a typical 3-Sylow subgroup is similar to $\mathrm{diag}(1, e^{2\pi i/3}, e^{-2\pi i/3})$, see for example [Th4]. In this case the virtual representation $\rho_+ - (1)$ (say) restricts to the invariant representation constructed in the theorem.

p-groups of rank 2, $p \geqslant 5$

We quote without proof the following classification theorem of B. Blackburn, see [B] or [Hp, III, Satz 12.4], where it is stated for *normal* elementary abelian subgroups.

Theorem 8.5

If $p \geqslant 5$ and G is non-abelian of rank 2 then G has a presentation of one of the following types:

(1) $\{A, B\colon A^{p^a} = 1, \quad B^{p^b} = A^{p^c}, \quad A^B = A^k, \quad c \geqslant 0, \quad k^{p^b} \equiv 1\,(p^a), \; p^c(k-1) \equiv 0\,(p^a)\}$

(2) $\{A, B, C\colon A^p = B^p = C^{p^{n-2}} = [A,C] = [B,C] = 1, \; [A,B] = C^{p^{n-3}}\}$

(3) $\{A, B, C\colon A^p = B^p = C^{p^{n-2}} = [B,C] = 1, \;\; [A, C^{-1}] = B, \; [B,A] = C^{sp^{n-3}}\}$.

where $n \geqslant 4$ and s equals either 1 *or some quadratic non-residue* mod p.

The methods of Chapters 4 and 6 show that the even-dimensional $\mathbb{Z}$-cohomology of a split metacyclic p-group is generated by Chern classes, and one seeks to generalise this result. Note first that a non-abelian group of order p^3 is of either Type(1) or Type(2), and that among the non-abelian groups of order p^4 ($p \geqslant 5$) listed in Appendix 3 below most are included in the list (8.5). Furthermore a group of Type(3) contains a subgroup of Type(2) of index p, generated by A, B and C^p.

One method for studying the even-dimensional cohomology of such a group is as follows. First determine the irreducible representations; for types 2 and 3 the group contains a normal abelian subgroup of index p, so that the degree of an irreducible representation is either 1 or p. Second determine a family of generators for the Chern ring, using a gloss on Theorem 6.3 to be explained below. Third consider the spectral sequence of a suitable group extension, which may not collapse, but for which knowledge of the Chern ring is enough to identify a family of universal cycles and hence generators for $H^{\text{even}}(G, \mathbb{Z})$. As an illustration of the method consider a group of Type(2) above.

Theorem 8.6

If G is a group of order p^n with a presentation of the form 8.5(2), *$H^{\text{even}}(G, \mathbb{Z})$ is generated by Chern classes.*

Proof. As an extension we shall need to look at G in two ways:

(i) $1 \to C^{C}_{p^{n-2}} \to G \to C^{\bar{A}}_{p} \times C^{\bar{B}}_{p} \to 1$, and

(ii) $1 \to C^{B}_{p} \times C^{C}_{p^{n-2}} \to G \rightleftarrows C^{A}_{p} \to 1$.

$$\underbrace{C^{B}_{p} \times C^{C}_{p^{n-2}}}_{\displaystyle \| \atop \displaystyle K}$$

The kernel in the second extension defines the normal abelian subgroup K of rank 2 and index p. Let $\hat{\gamma}$ be the one-dimensional representation of K which maps B to 1 and C to $e^{2\pi i/p^{n-2}}$. Following our usual convention write $\gamma = c_1(\hat{\gamma}) \in H^2(K, \mathbb{Z})$.

$$H^2(G, \mathbb{Z}) = (\mathbb{Z}/p)^{\alpha} \times (\mathbb{Z}/p)^{\beta} \times (\mathbb{Z}/p^{n-3})^{\tilde{\gamma}}$$

The commutator subgroup of G is generated by $C^{p^{n-3}}$, and we distinguish between the cohomology element $\tilde{\gamma}$, which is the first Chern class of the representation of G mapping A, B to 1 and C to $e^{2\pi i/p^{n-3}}$, and γ which is the first Chern class of the representation of K mapping B to 1 and C to $e^{2\pi i/p^{n-2}}$. The irreducible representations of G are either 1-dimensional, or of the form $i_! \hat{\gamma}$, where in the latter case there are $p^{n-3}(p-1)$ distinct primitive roots to choose as the image of C. Note that

$$p^n = p^2 p^{n-3} + p^2(p^{n-3}(p-1)).$$

Lemma 8.7

Let p be an odd prime, and let ρ be a one-dimensional representation of the normal subgroup K of G. Suppose that the index $[G{:}K] = p$, *and that* $[G{:}1] = p^n$. *Then*

$$s_k(i_! \rho) = i_*(s_k \rho) = i_*(c_1 \rho)^k, \quad \text{if } k < p-1 \text{ or } k = p,$$

and

$$s_{p-1}(i_! \rho) = i_*(s_{p-1}\rho) + (p-1)\operatorname{Inf}(\alpha^{p-1}), \quad \textit{where } \alpha \textit{ generates } H^2(G/K, \mathbb{Z}).$$

Proof. The first part follows from Theorem 6.3. For the second part embed G as a subgroup of the Wreath product $K \wr S_p$ as in Chapter 6, see pages 66–7. If K is normal in G, its image is contained in the normal subgroup $K \times \cdots \times K$ (p factors) of the Wreath product, and as already observed,

$$i_! \rho = i^*_{K \wr S_p \to G}(\rho \wr 1).$$

The characteristic classes of the transferred representation may now be calculated from those of $\rho \wr 1$, which in turn are detected on some page of the spectral sequence

$$E_2^{i,j} = H^i(S_p, H^j(K^p, \mathbb{Z})) \Rightarrow H^*(K \wr S_p, \mathbb{Z}).$$

In particular $s_k(i_! \rho)$ equals the sum of a term in $E_2^{0,k} = H^k(K^p, \mathbb{Z})^{S_p}$, a term in $E_2^{k,0} = H^k(S_p, \mathbb{Z})$ and various cross terms which depend on $c_j(\pi_l)$, $j < k$ and $l \leqslant p$. As in the previous chapter π_l denotes the permutation representation of the symmetric group S_l. The term on the fibre corresponds to $i_*(s_k(\rho \int 1))$, since this restricts to $(c_1 \rho)^k$ plus the sum of its conjugates under coset representatives $X_2, X_3, \ldots, X_p$ (say) by the double coset formula (3.1(ii)). The term on the base is clearly equal to $s_k(\pi_p)$. Moreover if $k = p-1$, on restriction to the sub-p-group G all the cross-terms vanish, since p does not divide the denominator of $B_{2j}/2j$ for $2j < k$, and $c_{2j+1}(\pi_l)$ is 2-torsion. The representation π_p restricts to the permutation representation of G on the set of cosets G/K, that is, factors through the regular representation of G/K. If α generates $H^2(G/K, \mathbb{Z})$, the total Chern class of the regular representation equals $1 - \alpha^{p-1}$. Using the recursive definition of the Newton polynomials and their naturality with respect to inflation we see that the restriction to G of the component in $E_2^{k,0}$ of $s_k(\rho \wr 1)$ equals $(p-1)\operatorname{Inf}(\alpha^{p-1})$. The lemma follows.

The description of the irreducible representations of the group G in the hypothesis of 8.6, together with 8.7 shows that the Chern subring of $H^{\text{even}}(G, \mathbb{Z})$ is generated by the family of elements described in the table below.

Symbol	α	β	$\tilde{\gamma}$	$i_*\gamma^l$	$\xi = c_p(i_!\tilde{\gamma})$
Dimension	2	2	2	$4 \leqslant 2l \leqslant 2p-2$	$2p$
Additive order	p	p	p^{n-3}	p^{n-2}	p^{n-1}

The additive orders of the higher dimensional generators again follow from the double coset rule. Thus

$$i^* i_*(\gamma^l) = p\gamma^l + \text{correction terms involving } \beta,$$

and $i_*(\gamma^l)$ has the maximum possible order p^{n-2}. The order of ξ equals p^{n-1}, since the expansion of $s_p(i_!\tilde{\gamma})$ in terms of Chern classes contains the term $p\xi$. The group G is another example for which the exponent is smaller than the exponent of the graded cohomology group.

It remains to show that $H^{\text{even}}(G, \mathbb{Z})$ contains no elements other than those expressible in terms of Chern classes. Consider the spectral sequence of the central extension (i) with terms

$$E_2^{i,j} \cong H^i(C_p \times C_p, H^j(C_{p^{n-2}}, \mathbb{Z})).$$

Since $H^{\text{odd}}(C_{p^{n-2}}, \mathbb{Z}) = 0$, the even dimensional cohomology is detected by subquotients of $E_2^{2i,2j}$, and since the extension is central

$$E_2^{0,2j} \cong H^{2j}(C_{p^{n-2}}, \mathbb{Z}),$$

generated, with a slight abuse of notation, by γ^j. It is clear that $p\gamma$ is a universal cycle corresponding to $\tilde{\gamma}$ in the cohomology of G. The vertical multiplication

$$\cup\gamma\colon E_2^{i,j} \to E_2^{i,j+2}$$

is a monomorphism for $j \geqslant 0$ (an isomorphism for $j > 0$), and the spectral sequence admits a horizontal multiplication

$$E_2^{i,2j} \circ E_2^{k,2j} \to E_2^{i+k,2j}, \quad j > 0,$$

defined by the product in the cohomology of $C_p \times C_p$ for fixed coefficients. In Figure 8.1, which describes page two near the origin, μ, $\nu \in E_2^{1,2}$ are independent generators, and $\chi = \mu \circ \nu$ in $E_2^{2,2}$. Since $H^2(G, \mathbb{Z})$ is of type (p, p, p^{n-3}), $d_3\gamma = s\delta$, $s \equiv 1 \pmod p$, and so

$$d_3\gamma^i = si\gamma^{i-1}\cdot\delta \neq 0, 1 \leqslant i \leqslant p,$$

by vertical periodicity. The generators of $H^4(C_p \times C_p, \mathbb{Z})$ are linearly independent, so $d_3\mu = d_3\nu = 0$, and since μ, ν come from the exterior subalgebra of $H^*(C_p \times C_p, \mathbb{F}_p)$ their powers make no contribution to $H^{\text{even}}(G, \mathbb{Z})$. The elements $\gamma^i\chi$ must all survive to infinity $(0 \leqslant i \leqslant p-2)$,

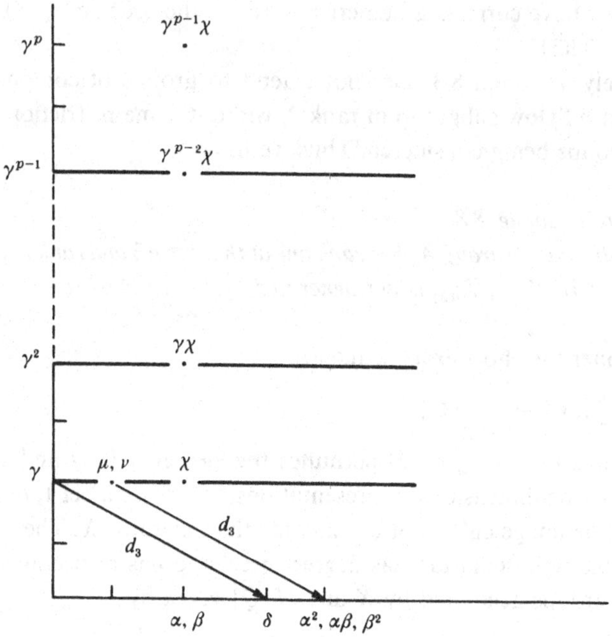

Figure 8.1

since the pair $(\gamma^i\chi, p\gamma^{i+1})$ is needed to detect the summand of order p^{n-2} generated by $i_*(\gamma^{i+1})$. This holds also in dimension $2p$, except that we must replace $p\gamma^p$ by γ^p (note that $d_3\gamma^p = 0$ above) in order to obtain the correct order for the top-dimensional Chern class, see [Le] for more information about this step. It is possible to check by hand that $H^4(G, \mathbb{Z})$ is generated as an abelian group by α^2, $\alpha\beta$, β^2, χ and $p\gamma^2$. However the entire spectral sequence becomes much easier to understand, if one compares it with that of the extension of the subgroup generated by $C^{p^{n-3}}$, A and B. This is studied exhaustively in [Le], and the conclusions for the cohomology of the group P_2 of order p^3 are listed for reference in Appendix 2.

On the later pages of the spectral sequence the vertical isomorphism $\cup\gamma$ is replaced by $\cup\chi\gamma\colon E_r^{2j,0} \to E_r^{2j+2,2i}$ for $j \geqslant 1$.

The calculation in dimension 4 together with the original vertical isomorphism shows that $E_3^{2,2i+2} = E_\infty^{2,2i+2}$ is generated by $\chi\gamma^i (i \geqslant 0)$. It follows that we have identified a family of universal cycles, and that $\mathrm{Ch}(G) = H^{\mathrm{even}}(G, \mathbb{Z})$.

For similar calculations for other rank 2 p-groups see [Al]. Note that the calculation above corrects a numerical error in the order of $s_k(i_!\hat{\gamma})$, as announced in [Th2].

Unfortunately Theorem 8.3 does not extend to groups of composite order having a p-Sylow subgroup of rank 2, without some restriction on the class of groups being considered. Thus we have

Counterexample 8.8

The alternating group A_4 *has rank one at the prime* 3 *and rank two at the prime* 2. *But* $H^{\text{even}}(A_4, \mathbb{Z})_{(2)}$ *is not generated by Chern classes.*

Proof. Consider the short exact sequence

$$1 \to C_2^P \times C_2^Q \to A_4 \to C_3^X \to 1,$$

in which conjugation in A_4 by X permutes the elements P, Q and PQ. There are three one-dimensional representations, which we label 1, ω, ω^2 depending on the image cube root of unity for the generator X. The only other irreducible representation has degree three, and has restrictions to the Sylow subgroups generated by X and $\{P, Q\}$ given by

$$1 + \omega + \omega^2$$

and by

$$\hat{p} + \hat{q} + (p \otimes q)\hat{}$$

respectively. Here $\hat{p}(P) = -1$, $\hat{p}(Q) = 1$ etc. The total Chern class

$$\begin{aligned} c.(\xi \mid A_{4,2}) &= (1 + p)(1 + q)(1 + p + q) \\ &= (1 + (p^2 + pq + q^2) + (p^2q + q^2p)). \end{aligned}$$

Since P and Q generate a normal subgroup of A_4, the image of $H^*(A_4, \mathbb{Z})_{(2)}$ equals the elements invariant under conjugation. In dimension six $H^*(A_{4,2}, \mathbb{Z})$ contains the monomials p^3, q^3, pq^2 and p^2q, and an easy calculation shows that $p^3 + pq^2 + q^3$, $p^3 + p^2q + q^3$ and $p^2q + q^2p$ are invariant. Only the third of these classes belongs to the Chern ring.

The alternating group A_4 is isomorphic to the projective special linear group $\mathrm{SL}(2, \mathbb{F}_3)$, and a similar argument shows that $H^*(\mathrm{PSL}(2, \mathbb{F}_q), \mathbb{Z})_{(2)}$ is not generated by Chern classes for $q \equiv \pm 3(8)$. The point of this restriction is that a 2-Sylow subgroup is still isomorphic to $C_2 \times C_2$; in the remaining cases it is dihedral and a more complicated numerical argument is required.

The heuristic reason for the good local result when the p-rank equals one is the single variable polynomial nature of the invariant elements in cohomology, except when $G_2 \cong D^*_{2^n}$. But here the period is 4 and one can use a special low-dimensional argument. Indeed there is an epimorphism

$$c_2 : R_4^\gamma(G) \twoheadrightarrow H^4(G, \mathbb{Z})_2,$$

corresponding to the isomorphism

$$c_1 : \mathrm{Hom}(G, \mathbb{C}^\times) \rightarrowtail\!\!\twoheadrightarrow H^2(G, \mathbb{Z}).$$

Here $R_4^\gamma(G)$ is the subgroup of $R(G)$ generated by the monomials

$$\Lambda^i(\rho_j - \varepsilon(\rho_j) + i - 1),$$

where $i \geqslant 2$ and $\rho_1, \ldots, \rho_j, \ldots, \rho_t$ are the irreducible representations. See the Appendix to [At] for more details.

Notes and references

This chapter is a reworking of several of my earlier papers. I came across the global version of 8.3 for groups with periodic cohomology while working on the spherical space form problem [Th1]. The local version presented here is I believe new, although there is an obvious link to the work in [J-Z]. The calculations for the rank 2 groups owe a great deal to G. Lewis [Le], who calculated $H^*(G, \mathbb{Z})$ completely for groups of order p^3. This work deserves to be more widely known – since in tandem with the information given by the formula in Theorem 6.3, it suggests numerous further calculations. Furthermore [Le] is really essential reading for anyone who wishes to flesh out the rather sketchy outline proof given for Theorem 8.6. The significance of the counterexample, 8.8, is explained in [Th6]; indeed one class of groups for which there is a local rank 2 theorem along the lines of 8.6 is the one for which the 'topological' and 'γ-filtrations' on the representation ring coincide.

9

Linear groups over finite fields

The special linear groups of 2×2 matrices over the finite field $\mathbb{F}_q$ ($q = p^t$ and p odd) provide a beautiful illustration of Theorem 8.3, and the calculations give some information about $\mathrm{SL}(n, q)$, $n \geqslant 3$. Furthermore, replacing $\mathbb{F}_q$ by a suitable approximation to the algebraic closure $\mathbb{F}_{q^\infty}$ one can develop a theory of universal Chern classes for modular representations of a finite group. This illustrates A. Grothendieck's theory sketched on pages 87–88 above.

SL(*n*, *q*)

Suppose first that $n = 2$. The order of $G = \mathrm{SL}(2, q)$ is $q(q^2 - 1)$, and the different conjugacy classes are represented by elements of the form:

(i) $\pm 1_2$,

(ii) diagonal matrices $Z = \begin{pmatrix} \alpha & 0 \\ 0 & \alpha^{-1} \end{pmatrix}$, $\alpha \in \mathbb{F}_q^\times$,

(iii) matrices W diagonalisable over $\mathbb{F}_{q^2}$ with eigenvalues γ, γ^{-1} such that $\gamma^{1+q} = 1$, and

(iv) lower triangular matrices of the form $Y = \begin{pmatrix} 1 & 0 \\ \beta & 1 \end{pmatrix}$. Write $X = \begin{pmatrix} 1 & 0 \\ 1 & 1 \end{pmatrix}$. Particularly important for the representation theory is the subgroup

$$B = \left\{ \begin{pmatrix} \alpha & 0 \\ \beta & \alpha^{-1} \end{pmatrix} : \alpha \in \mathbb{F}_q^\times, \beta \in \mathbb{F}_q \right\}$$

of index $(q + 1)$, which contains the normal cyclic subgroup

$$U = \left\{ \begin{pmatrix} 1 & 0 \\ \beta & 1 \end{pmatrix} : \beta \in \mathbb{F}_q \right\}$$

of unipotent lower triangular matrices with composition defined by addition in $\mathbb{F}_q$. The quotient B/U is isomorphic to $D=\left\{\begin{pmatrix}\alpha & 0\\ 0 & \alpha^{-1}\end{pmatrix}: \alpha\in\mathbb{F}_q^\times\right\}$, which is cyclic of order $q-1$ and which operates on $\mathbb{F}_q$ by the square of the usual action. The normaliser of the abelian subgroup generated by elements of type (iv) equals B. For type (ii) the normaliser is contained in an extension of D by the subgroup generated by $\begin{pmatrix}0 & 1\\ -1 & 0\end{pmatrix}$, and for type (iii) in the product of $\mathbb{F}_{q^2}^\times$ by an element of order four inducing the Frobenius automorphism $\delta\longmapsto\delta^q$. From this one concludes that $H^*(G,\mathbb{Z})_l$ has period equal to 4 for all primes $l\neq p$, the characteristic of $\mathbb{F}_q$ – use the normaliser–centraliser condition 3.4, or 8.4 if $l=2$. A p-Sylow subgroup is isomorphic to U, elementary abelian of rank t, the dimension of $\mathbb{F}_q$ as a vector space over $\mathbb{F}_p$. In particular if $t=1$, $H^*(G,\mathbb{Z})_{(p)}$ has period equal to $(q-1)$. Note that $2[N(U)\colon Z(U)]=(p-1)$ in this limiting case, when $p=q$.

Remark

Since G is perfect, $H^2(G,\mathbb{Z})=0$ and $H^{4k+2}(G,\mathbb{Z})_{(2)}=0$ by 2-periodicity. All this can be summarised as

Theorem 9.1

$H^*(\mathrm{SL}(2,q), Z[1/p])$ is a polynomial ring over $\mathbb{Z}/(q^2-1)$ on a 4-dimensional generator. If $p=q$, $H^*(\mathrm{SL}(2,p),\mathbb{Z})_{(p)}$ is also polynomial over $\mathbb{Z}/p$ on a $(p-1)$-dimensional generator.

The group G has the character table shown in Table 9.1 see for example [Sc].

Table 9.1

Conjugacy classes	1	1	$q-3/2$	$q-1/2$	2	2
1_2	1	q	$q+1$	$q-1$	$\frac{1}{2}(q+1)$	$\frac{1}{2}(q-1)$
-1_2	1	q	$(-1)^u(q+1)$	$(-1)^v(q-1)$	$\frac{\varepsilon}{2}(q+1)$	$-\frac{\varepsilon}{2}(q-1)$
X	1	0	1	-1	$\frac{1}{2}(1\pm\sqrt{(\varepsilon q)})$	$\frac{1}{2}(-1\pm\sqrt{(\varepsilon q)})$
Y	1	0	1	-1	$\frac{1}{2}(1\mp\sqrt{(\varepsilon q)})$	$\frac{1}{2}(-1\mp\sqrt{(\varepsilon q)})$
Z^a	1	1	$\zeta^{ua}+\bar{\zeta}^{ua}$	0	$(-1)^a$	0
W^b	1	-1	0	$-(\omega^{vb}+\bar{\omega}^{vb})$	0	$-(-1)^b$

Here $1\leqslant a,\ u\leqslant(q-3)/2,\ 1\leqslant b,\ v\leqslant(q-1)/2,\ \varepsilon=(-1)^{(q-1)/2}$, and ζ (respectively ω) is a primitive $(q-1)$st root (respectively $(q+1)$st root) of unity.

If α generates $\mathbb{F}_q^\times$, let $\hat{\alpha}_u$ be the 1-dimensional representation of B which maps the matrix $\begin{pmatrix} \alpha & 0 \\ \beta & \alpha^{-1} \end{pmatrix}$ to ζ^u. Then the transferred representation $i_! \hat{\alpha}_u$ has degree $(q+1)$ and supplies the entries in column three for $1 \leqslant u \leqslant (q-3)/2$. If $u=0$, $i_! \hat{\alpha}_0 = 1 + \rho$, with degree $\rho = q$ (column 2). If $u = (q-1)/2$, $i_! \hat{\alpha}_u = \xi_+ + \xi_-$, giving the entries in column five. Let $\hat{\beta}_v$ map W to ω^v; then $\hat{\theta}_v$ is a well-defined representation with character

$$\chi_\rho \chi_{i_! \hat{\alpha}_v} - \chi_{i_! \hat{\alpha}_v} - \chi_{i_! \hat{\beta}_v}.$$

The representation $\hat{\theta}_v$ is irreducible unless $v = (q+1)/2$; $\hat{\theta}_{(q+1)/2}$ splits as $\eta_+ + \eta_-$. These representations (columns 4 and 6) are called cuspidal.

Suppose that $p = q$, so that G has periodic cohomology for all primes dividing the order. For the two exceptional cuspidal representations $\eta_\pm$,

$$\chi(X) = \chi\begin{pmatrix} 1 & 0 \\ 1 & 1 \end{pmatrix} = \tfrac{1}{2}\left(-1 \pm \sqrt{\left[(-1)^{\frac{q-1}{2}} q\right]}\right),$$

and restricted to the cyclic subgroup generated by X of order p the sum $\eta_+ + \eta_-$ has eigenvalues equal to the primitive pth roots of unity, each counted once. Half the primitive roots correspond to η_+ (say), and from the properties of the Chern classes given in Chapter 6.

$$c_{(q-1)/2}(i_{!G\to U}(\eta_+)) \text{ generates } H^{q-1}(U, \mathbb{Z}).$$

Since the normaliser–centraliser argument shows that $H^k(G,\mathbb{Z})_{(p)} \cong H^k(U,\mathbb{Z})$ for $k \equiv 0 \,(\mathrm{mod}\,(p-1))$ $c_{(q-1)/2}(\eta_+)_{(p)}$ generates the p-torsion subring in the cohomology of G. For $l \neq p$, and removing the restriction $p = q$, if l divides $q-1$, we may use powers of $c_2(i_! \hat{\alpha}_1)$ to generate $H^{4k}(G,\mathbb{Z})_{(l)}$, and if l divides $q+1$ powers of $c_2(\hat{\theta}_1)$ (see the table above). This illustrates the working of Theorem 8.3 for one particular family of groups.

Exercise

Let $t = 2$, so that $G_p \cong C_p \times C_p$ and $rk_p G = 2$. Since G_p is abelian, Swan's Lemma (3.4) implies that the stable elements in $H^*(C_p \times C_p, \mathbb{Z})$ equal the elements invariant under the action of the normaliser B. Calculate these invariant elements, and compare the result with the p-primary part of the Chern ring, obtainable from the character table.

The cohomology of the general linear group $\mathrm{GL}(2,q)$ is more complicated, but may be studied by means of the spectral sequence of the extension

$$\mathrm{SL}(2,q) \rightarrowtail \mathrm{GL}(2,q) \twoheadrightarrow \mathbb{Z}/q-1 \cong \mathbb{F}_q^\times,$$

and the known character table, see for example the last page of [P – Sh]. The subgroup D of diagonal matrices is now isomorphic to $\mathbb{F}_q^\times \times \mathbb{F}_q^\times$, and for each odd prime number l dividing $(q-1)$ the l-rank equals two. Restriction maps $H^*(\mathrm{GL}(2,q), \mathbb{Z})_{(q-1)}$ monomorphically into $H^*(D, \mathbb{Z})$, the structure of which is given in 8.1. The even-dimensional subring has a 2-dimensional generator (detected by $E_2^{2,0}$), and a 4-dimensional generator (detected by $E_2^{0,4}$). The former is a first Chern class; the latter a second Chern class of a representation constructed in the same way as $i_! \hat{\alpha}_u$ above, starting with the character

$$\begin{pmatrix} \alpha_1 & 0 \\ \beta & \alpha_2 \end{pmatrix} \longmapsto \zeta_1^{u_1} \zeta_2^{u_2},$$

ζ_1 and ζ_2 being two copies of the root ζ, corresponding to the two copies of $\mathbb{F}_q^\times$ in D. The $(q+1)$-torsion is easier to handle, since the only prime to divide both $(q-1)$ and $(q+1)$ is 2. Therefore, if l divides $q+1$ and l is odd, the spectral sequence with local coefficients $\mathbb{Z}_{(l)}$ is trivial, and

$$H^*(\mathrm{GL}(2,q), \mathbb{Z}_{(l)}) \cong H^*(\mathrm{SL}(2,q), \mathbb{Z}_{(l)})^{\mathrm{inv}}$$

As in the case of the special linear group one calculates l-torsion using powers of the second Chern class of a cuspidal representation, see [P – S] *op. cit. supra.* Hence

Theorem 9.2

If $G = \mathrm{GL}(2,q)$, the Chern ring Ch(G) *localised away from the primes p and 2 is isomorphic to $H^{\mathrm{even}}(G, \mathbb{Z}[\frac{1}{2}, 1/p])$.*

Suppose now that $p \geqslant 3$, $n \geqslant 3$. From [Hp] we quote the following results on the subgroup structure of $\mathrm{SL}(n,q)$:

(1) The order of $\mathrm{SL}(n,q)$ equals $(q^n-1)(q^n-q)\ldots(q^n-q^{n-2})q^{n-1}$, (II 6.2).

(2) The subgroup of diagonal matrices D has rank $n-1$, order $(q-1)^{n-1}$, and is normalised by a subgroup of the symmetric group S_n, (II 7.2).

(3) The group $\mathrm{SL}(n,q)$ contains a cyclic subgroup L of order $(q^n-1)/(q-1)$, generated by the so-called Singer cycle. L is its own centraliser and $N(L)/L$ is cyclic of order n. Furthermore, if $L' \subseteq L$ and the order of L' is not a proper divisor of $q^m - 1$ for all $m < n$, then L' and L have the same normaliser, (II 7.3).

It follows from (3) and the normaliser–centraliser condition that if l is an odd prime dividing $q^n - 1$, but no $q^m - 1$ for $1 \leqslant m < n$, then $H^*(\mathrm{SL}(n,q), \mathbb{Z}_l)$ is a polynomial ring with a single generator in dimension $2n$. This generator may be interpreted as the nth Chern class of the virtual complex representation obtained by 'lifting' the identity map $\mathrm{SL}(n,q) \to \mathrm{SL}(n,q)$ to characteristic zero via some embedding of $\mathbb{F}_q^\times$ in the complex numbers $\mathbb{C}$ (see also the following section and [Gr]).

As a very special case take $n=3$ and $p=q=$ odd. Then $\mathrm{SL}(3,q)$ has order equal to $q^3(q^2-1)(q^3-1)$; if l divides $(q+1)(q^2+q+1)$ and $l \geqslant 5$ the corresponding Sylow subgroup is cyclic. If $l=2,3,q$ or l divides $q-1$, and we restrict attention to the case $q \equiv 3(\mathrm{mod}(4))$, then the l-rank equals 2. In particular a q-Sylow subgroup is of elementary type P_2 and a 2-Sylow subgroup is semidihedral, see [Hp, page 190]. If $L' \subseteq L$ has order $l(\geqslant 5)$, then l divides neither $q+1$ nor $q-1$, so by (3) above the subgroups L' and L have the same normaliser. Therefore the l-period of $H^*(\mathrm{SL}(3,q),\mathbb{Z})$ equals 6, and c_3 of the representation just described defines a ring generator. More generally one may apply Theorem 8.3 to $\mathrm{SL}(3,q)$ for $l \geqslant 5$, l not a divisor of $q(q-1)$, and state

Theorem 9.3

If $l \geqslant 5$, $l \nmid q(q-1)$, *then*

$$\mathrm{Ch}(\mathrm{SL}(3,q))_{(l)} = H^*(\mathrm{SL}(3,q),\mathbb{Z})_{(l)}$$

is polynomial on either a 6-dimensional $(l|q^2+q+1)$ *or a 4-dimensional generator* $(l|q+1)$

Note that if l divides $q+1$ one may choose a representative l-Sylow subgroup to lie inside the subgroup $\mathrm{SL}(2,q)$. If l divides $q-1$ there is a similar result (see the exercise following Theorem 9.6), but one needs the methods of the next section.

Characteristic classes for modular representations

Let l be a prime number. In contrast to the previous section suppose that l divides $q-1$. We start with a specific model for the classifying space of the Wreath product $G \wr C_l$, where C_l acts on $G \times \cdots \times G$ (l factors) by permutation. Recall from Chapter 6 that $G \wr C_l$ also acts on the contractable product of classifying spaces

$$(EG)^l \times EC_l,$$

and that the orbit space $(BG)^l \times_{C_l} EC_l$ is a model for the classifying space of the Wreath product. The inclusion maps $i: G^l \to G \wr C_l$ and $j: G \times G_l \to G \wr C_l$, the latter being defined by the product of the diagonal and of a splitting map, induce maps of classifying spaces

$$\underbrace{BG \times \cdots \times BG}_{l} \xrightarrow[Bi]{} (BG)^l \times_{C_l} EC_l \xleftarrow[Bj]{} BG \times BC_l$$

Lemma 9.4
The induced map in cohomology

$$H^*(G \wr C_l, \mathbb{F}_l) \xrightarrow[(i^*, j^*)]{} H^*(G^l, \mathbb{F}_l) \oplus H^*(G \times C_l, \mathbb{F}_l)$$

is a monomorphism.

Proof. If $v \in H^2(C_l, \mathbb{F}_l)$ is a generator for the polynomial part of this cohomology ring, then on inverting v the map j^* becomes an isomorphism, thus:

$$H^*(G \wr C_l, \mathbb{F}_l)[v^{-1}] \xrightarrow{\cong} H^*(G \times C_l, \mathbb{F}_l)[v^{-1}].$$

This is a consequence of a localisation theorem in equivariant cohomology – if X is an arbitrary compact C_l-space with a given finite triangulation or cellular decomposition, and $H^*_{eq}X = H^*(X \times_{C_l} EC_l, \mathbb{F}_l)$, then

$$H^*_{eq}(X)[v^{-1}] \cong H^*_{eq}(F)[v^{-1}],$$

F being the fixed point set of the action. (The idea of the proof is as follows: if F is empty, since l is prime, the action on X is free and $H^*_{eq}(X) \cong H^*(X/C_l, \mathbb{F}_l)$. However, since X is compact and hence finite dimensional, pulling back along the equivariant map $X \to$ point shows that $H^*(X/C_l, \mathbb{F}_l)$ is annihilated by some power of v, proving the claim in this case. More generally one applies a Mayer–Vietoris exact sequence argument to the triple $(X - \mathring{N}, N, N \cap \mathring{N})$, where N is an equivariant regular neighbourhood of F, and then passes to the limit over a contracting family of such neighbourhoods, obtained by subdivision, say, of the original triangulation.) The finiteness assumption on X does not matter for the present application, since we may replace each classifying space by a finite skeleton, and observe that with $\mathbb{F}_l$-coefficients cohomology transforms direct limits into good inverse limits.

So far one knows that any element in the kernel of j^* is killed by a power of v. Consider the spectral sequence of the defining exact sequence of the Wreath product with $\mathbb{F}_l$-coefficients:

$$E_2^{i,j} = H^i(C_l, H^j(G^l, \mathbb{F}_l)).$$

A non-zero element y of $H^*(G \wr C_l, \mathbb{F}_l)$ lying in the kernel of i^* is represented by an element z in $E_2^{i,j}$ for some $i > 0$. Recall at this point that the fibre terms in the spectral sequence describe the image of the restriction map – in the present case, by using the chains in the chosen model for $B(G \wr C_l)$, one sees that the image of i^* coincides with the invariant elements, see [N, page 238] for more details. Hence $E_2^{0,j} = E_\infty^{0,j}$. Periodicity implies that taking products with the universal cycle v^k from the base defines an isomorphism.

$$E_2^{i,j} \cong E_2^{i+2k,j},$$

So z is not killed by any power of v. If y were killed by the power v^k, zv^k would have to be hit from behind by some differential. If this were away from the fibre, z would also have to be hit, and we have already noted that all differentials are trivial on the fibre. This shows that the intersection of the kernels of i^* and j^* is zero, proving the lemma.

Remark

There is actually an isomorphism $H^*(G \wr C_l \mathbb{F}_l) \cong H^*(C_l, H^*(G^l, \mathbb{F}_l))$, see [N], Theorem 3.3.

We will say that the pair of subgroups G^l and $G \times C_l$ *detects* the mod l cohomology of $G \wr C_l$. More generally, if $\{A_i : i \in I\}$ is a detecting family of subgroups for G, then the family $\{A_{i_1} \times \cdots \times A_{i_k}; A_{i_k} \times C_l : i_1, \ldots, i_l, i_k \in I\}$ is a detecting family for $G \wr C_l$.

Corollary

Let G be a group whose mod 1 cohomology is detected by a family of abelian subgroups of exponent dividing l^a with $a \geqslant 1$. Then $G \wr S_n$ has the same property.

Proof. Assume that n is greater than one.

Case 1: n is not divisible by l. The subgroup $(G \wr S_{n-1}) \times G$ detects mod l cohomology because its index is coprime with l, and the assertion follows by induction.

Case 2: $n = n'l$. Then $[G \wr S_n : G \wr C_l) \wr S_{n'}]$ equals $n/n'!l^{n'}$, which is seen not to be divisible by l. The argument above plus induction gives a detecting family of a abelian subgroups of exponent dividing l^a for the subgroup $(G \wr C_l) \wr S_{n'}$, which is good enough.

Assertion (2) for the special linear group on page 105 above applies to the general linear group over the finite field $\mathbb{F}_q$, and if D is the diagonal subgroup, one has

$$N_{GL(n,q)}(D) \cong \mathbb{F}_q^\times \wr S_n$$

Theorem 9.5

The mod l cohomology of $\mathrm{GL}(n, q)$ *is detected by the subgroup D of diagonal matrices, if l divides $(q-1)$ or if $l = 2$ and 4 divides $(q-1)$.*

In this theorem the prime power q need not be odd.

Proof. The order of $\mathrm{GL}(n, q)$ equals $(q^n - 1)(q^n - q)\cdots(q^n - q^{n-1})$; that of the subgroup D equals $(q-1)^n n!$. The index therefore equals

$$q^{n(n-1)/2} \cdot \frac{q^n - 1}{n(q-1)} \cdot \frac{q^{n-1} - 1}{(n-1)(q-1)} \cdots \frac{q^2 - 1}{2(q-1)},$$

each factor of which is an l-adic unit under the hypotheses on l, and so the index is coprime with l. By the corollary to the last lemma $\mathrm{GL}(n, q)$ has a detecting family for mod l cohomology, which consists of abelian groups A a exponent dividing $q - 1$. By elementary linear algebra A is conjugate to a subgroup of the maximal torus

$$D = \underbrace{\mathbb{F}_q^\times \times \cdots \times \mathbb{F}_q^\times}_{n}$$

Put another way, the representation module $(\mathbb{F}_q^n, A)$ splits into a sum of 1-dimensional submodules, since the exponent of A (equal to a power of l) divides $q - 1$.

Let $\bar{\mathbb{F}}_p$ be an algebraic closure of $\mathbb{F}_p$ and let $\phi: \bar{\mathbb{F}}_p^\times \to \mathbb{C}^\times$ be some embedding. We have already noted that the standard representation of $\mathrm{GL}(n, \mathbb{F}_q)$ in $\mathrm{GL}(n, \bar{\mathbb{F}}_q)$ lifts, using ϕ, to a virtual complex representation, and hence, using the flat bundle construction, to a map of classifying spaces $B\mathrm{GL}(n, q) \to B\mathrm{GL}$. The maps obtained in this way are compatible as $n \to \infty$ and as $\mathbb{F}_q$ becomes larger. Hence for any subfield $\mathfrak{k}$ of $\bar{\mathbb{F}}_p$ we obtain an element $\hat{\sigma} \in \lim_{\mathbb{F}_q \subset \mathfrak{k}} R(\mathrm{GL}(n, q))$, and a map

$$B\mathrm{GL}(n, \mathfrak{k}) \to B\mathrm{GL},$$

which is well-defined up to homotopy. Let $c_k(\hat{\sigma})$, $1 \leqslant k \leqslant n$, be the image of the reduction modulo l^a of the universal Chern class, defined in the last section of Chapter 5, under this map in $H^{2k}(B\mathrm{GL}(n, \mathfrak{k}), \mathbb{Z}/l^a)$.

Theorem 9.6.

Let l be a prime number distinct from p, the characteristic of $\mathfrak{k}$, and assume that $\mathfrak{k}$ contains the l^b-th roots of unity in $\bar{\mathbb{F}}_p$ for all b. Then

$$H^{\mathrm{even}}(B\mathrm{GL}(n, \mathfrak{k}), \mathbb{Z}/l^a) = \mathbb{Z}/l^a[c_1\hat{\sigma}, \ldots, c_n\hat{\sigma}].$$

Proof. We may without loss of generality suppose that $a = 1$, because the classes c_k are even dimensional and the relevant coboundary homomorphisms vanish. Consider the restriction homomorphism

$$H^{\text{even}}(B\mathrm{GL}(n, \mathfrak{k}), \mathbb{F}_l) \to H^{\text{even}}(BD_n(\mathfrak{k}), \mathbb{F}_l)^{S_n},$$

where the image is contained in the subgroup let invariant by the action of S_n (the Weyl group). The multiplicative group $\mathfrak{k}^\times$ is an increasing union of cyclic groups, and is l-divisible by hypothesis; hence passing to the limit over this union, and once more appealing to the fact that we use finite coefficients,

$$H^{\text{even}}(B\mathfrak{k}^\times, \mathbb{F}_l) = \mathbb{F}_l[\xi],$$

Where ξ is the first Chern class of the representation described by the embedding ϕ. By the Künneth formula for coefficients in a field

$$H^{\text{even}}(BD(\mathfrak{k}), \mathbb{F}_l) = \mathbb{F}_l[\xi_1, \ldots, \xi_n],$$

Where ξ_k is the image of ξ under the kth projection map. The group S_n acts by permutation, so (compare Chapter 5 again) the even dimensional invariant elements are the symmetric functions in the ξ_k $(1 \leqslant k \leqslant n)$. The restriction to $c_j(\hat{\sigma})$ to $D(\mathfrak{k})$ is the jth elementary symmetric function in the ξ_k – indeed using characters one sees that the restriction of $\hat{\sigma}$ to $D(\mathfrak{k})$ splits as the direct sum of the one-dimensional representations with first Chern classes equal to $\xi_1, \ldots, \xi_n$. Hence the product formula for the total Chern class gives the jth elementary symmetric function. Therefore the restriction map in 9.6 is an epimorphism; by 9.5 it is a monomorphism also.

Exercise

Let $G = \mathrm{SL}(3, \mathbb{F}_p)$; if l divides $p - 1$, an l-Sylow subgroup is isomorphic to $C_l \times C_l$ and hence has rank 2. Since $H^{\text{even}}(G_l, \mathbb{Z})$ is generated by classes which project to the polynomial part of $H^*(G_l, \mathbb{F}_l)$, see 8.1, one can use the arguments of 9.5 and 9.6 to describe $H^{\text{even}}(G, \mathbb{Z})_{(l)}$. Is this subring generated by Chern classes? Compare the answer for this group with 8.3 and 8.8.

Remarks on the Pontrjagin and Stiefel–Whitney classes

With the obvious changes in notation one can prove that if the characteristic of $\mathfrak{k}$ is odd, and $\mathfrak{k}$ contains $\sqrt{-1}$, then

$$H^{\text{even}}(B\mathrm{O}(n, \mathfrak{k}), \mathbb{F}_l) = \begin{cases} (l = \text{odd}) & \mathbb{F}_l[p_1, \ldots, p_m], m = [n]/2, \\ (l = 2) & \mathbb{F}_2[w_1, \ldots, w_n]. \end{cases}$$

See [Q1] for details – and note that throughout Quillen arranges his proofs with the Pontrjagin rather than the Chern classes in mind. As in the case of the general linear group explicit examples show that restriction into the invariant cohomology of a maximal torus (with Weyl group $C_2 \wr S_m$ or S_n) is an epimorphism. A variant of 9.5 then proves that restriction is (1–1).

In the final section of Chapter 7 we sketched how it is possible, using the language of Grothendieck topologies, to construct characteristic classes with finite or l-adic coefficients for representations of any discrete group in an algebraically closed field (char $(\mathfrak{f}) \neq l$). For representations in the field $\bar{\mathbb{F}}_p$ (modular representations) the last theorem suggests an alternative method of construction. We start with the universal classes

$$c_k(\hat{\sigma}) \in H^{2k}(BGL(n, \bar{\mathbb{F}}_p), \mathbb{Z}_l), \quad l \neq p,$$

defined by a compatible family of l^a-classes with $\mathfrak{f} = \bar{\mathbb{F}}_p$.

Definition

If $\rho: G \to GL(n,p) \to GL(n, \bar{\mathbb{F}}_p)$ is a modular representation of the finite group G, then

$$c_k(\rho) = (B\rho)^* c_k(\hat{\sigma}) \in H^{2k}(G, \mathbb{Z}_l), 1 \leqslant k \leqslant n.$$

Theorem 9.7

The classes just defined satisfy the properties (1)–(3) *stated in* 5.10. *Property* (4) *is replaced by*

(4)′ *if ρ is one-dimensional, $c_1(\rho)$ equals the image of the class of ρ in $H^1(G, \mathfrak{f}^\times)$ under the coboundary homomorphism*

$$\beta: H'(G, \bar{\mathbb{F}}_p^\times) \to H^2(G, \mu_{l^a}).$$

(Here as in Chapter 7 μ_{l^a} is the cyclic group of roots of unity, non-canonically isomorphic to $\mathbb{Z}/l^a$).

Sketch proof for the reader familiar with classical bundle theory. The dimension property and the naturality of the classes are both obvious from the definitions. For the exponential formula for the total Chern class recall that if ρ is a representation of G in $GL(n, \mathbb{F}_q)$, and $E(n)$ is the universal $GL(n, \mathbb{C})$-bundle over $BGL(n, \mathbb{C})$, then $c.(\rho)$ is the pull-back in cohomology of the class $c.E(n)$ along the composition of maps

$$BG \underset{B\rho}{\to} BGL(n, \mathbb{F}_q) \underset{B_i}{\to} BGL(n, \bar{\mathbb{F}}_p) \underset{B\hat{\sigma}}{\to} BGL(n, \mathbb{C}).$$

The uniqueness up to bundle isomorphism of the universal bundle $E(n)$ implies that over $BGL(n_1, \mathbb{C}) \times BGL(n_2, \mathbb{C})$ the restriction of $E(n_1 + n_2)$ is equivalent to the (exterior) product of $E(n_1)$ and $E(n_2)$. By choosing compatible inverse limit representations $\hat{\sigma}$ for increasing values of n it follows that the equation

$$c.(\text{Res}\, F(n_1+n_2)) = c.(E(n_1))c.(E(n_2))$$

also holds for the universal modular classes. Property (3) is now clear by naturality. For property (4)′ consider the universal 1-dimensional representation

$$1_{\bar{\mathbb{F}}_p}: \bar{\mathbb{F}}_p^\times \to \bar{\mathbb{F}}_p^\times.$$

Then $c_1(1_{\bar{\mathbb{F}}_p^\times}) = c_1(1_{\mathbb{C}^\times}\phi) = \phi^*\beta(1_{\mathbb{C}^\times})$, which coincides with $\beta(1_{\bar{\mathbb{F}}_p^\times})$, assuming that ϕ^* identifies the roots of unity in the two algebraically closed fields concerned. Here $1_{\mathbb{C}^\times}$ and $1_{\bar{\mathbb{F}}_p^\times}$ represent cohomology classes.

The question now arises as to the extent to which properties (1)–(4), determine the classes which we have defined using Theorem 9.6 and the classical topological theory of complex vector bundles. For convenience restrict attention to finite coefficients and note that for each prime $l \neq p$ we may confine attention to the restriction of ρ to an l-Sylow subgroup G_l of G. Thus

$$\rho_l: G_l \to \mathrm{GL}(n, \bar{\mathbb{F}}_p).$$

Since $p \neq l$ $\bar{\mathbb{F}}_p(G_l)$ is a semisimple algebra over an algebraically closed field, and, if G_l is abelian, any representation module splits as a sum of irreducible one-dimensional submodules. We have

Theorem 9.8

Let G be a finite-group such that the mod l cohomology of a representative l-Sylow subgroup G_l is detected by elementary abelian subgroups A. If c_k and c'_k are two families of (mod l) characteristic classes satisfying (1)–(4)′ for all representations ρ of G in $\mathrm{GL}(n, \bar{\mathbb{F}}_p)$, *then*

$$c_k(\rho) = c'_k(\rho).$$

Proof. Properties (1)–(4)′ imply that $c_k = c'_k$ for sums of one-dimensional representations, and hence for each elementary abelian subgroup A. We assume that the class of such subgroups detects the mod l cohomology of G_l, and the result follows.

Lemma 9.4 and its corollary show that the symmetric group S_n satisfies the hypothesis of 9.8. More generally it is known that the restriction of

$H^*(G_l, \mathbb{F}_l)$ to the direct sum of the cohomologies of its elementary abelian subgroups $H^*(A, \mathbb{F}_l)$ has nilpotent kernel. Hence, using simple topological methods, the Chern classes of modular representations are uniquely defined up to nilpotent elements, compare [Q2]. Note that the discussion of uniqueness, as opposed to that of existence, is independent of bundle theory. As a final result we quote another theorem of D. Quillen which shows that there is no chance of a theory of p-adic characteristic classes for modular representations in characteristic p, see [Q4] Theorem 6.

Theorem 9.9
$H^k(\mathrm{GL}(n, \mathbb{F}_{p^t}), \mathbb{F}_p) = 0$ *for* $0 < k < t(p-1)$ *and for all* n.

From this it follows by a limiting argument that if $\mathfrak{k}$ is an infinite algebraic extension of $\mathbb{F}_p$, then

$$H^k(B\mathrm{GL}(n, \mathfrak{k}), \mathbb{F}_p) = 0 \text{ for all } k > 0.$$

Quillen's method of proof is to show that

$$H^k(U, \mathbb{F}_p)^D = 0 \quad \text{for } 0 < k < t(p-1),$$

where U is the subgroup of upper triangular matrices with ones along the diagonal, and where the subgroup D of diagonal matrices normalises U. The methods used in the first half of this chapter can be used to illustrate the analogue of this theorem for the special linear groups ($t = 1$, $n \leqslant 3$) – compare the calculation of $H^{p-1}(\mathrm{SL}(2, p), \mathbb{Z})_{(p)}$ in 9.1. This calculation extends to $n = 3$; we need the D-invariant subring in $H^*(P_2, \mathbb{Z})$, listed in Appendix 2. Here as usual P_2 has order p^3 and exponent p.

Notes and references

As with Theorem 8.3 the non-stable calculations for the integral cohomology of $\mathrm{SL}(n, q)$ in the first part of this chapter were motivated by work on the topological space form problem. Indeed the calculation of $\mathrm{Ch}(\mathrm{SL}(2, p))_{(p)}$ first suggested the local theorem (8.3) for rank one groups. In the second part I have followed the relevant parts of D. Quillen's paper on the Adams Conjecture [Q1] – it is very interesting how, using elementary topological ideas, Theorem 9.6 confirms the existence of l-adic characteristic classes for representations in the algebraically closed field $\bar{\mathbb{F}}_p$. A possible development might be to examine the relation between the more delicate calculations of $H^*(\mathrm{GL}(n, \mathbb{F}_q), \mathbb{F}_{l^a})$ in [Q4] and the 'mixed' characteristic classes of Grothendiek [G] for representations over a not necessarily algebraically closed field. These are the classes already mentioned in the notes to Chapter 7, see [So].

Appendix 1
The Riemann–Roch formula

Using a delicate calculation from algebraic topology we shall outline a proof of Theorem 6.3 with the bound $\bar{M}_k$ replaced by M_k. Thus

Theorem

Let $M_k = (\prod_{\substack{p \leqslant k+1 \\ p \text{ prime}}} p^{[k/p-1]})/k!$, *let* ρ *be a representation of the subgroup* K *of the finite group* G, *and let* $s_k(\rho) = N(c_1(\rho)\dots c_k(\rho))$. *Then* $M_k(s_k(i_!\rho) - i_*(s_k\rho)) = 0$.

We have the following important special case (compare Lemma 8.7): *Let* ρ *be a one-dimensional representation of the normal subgroup* K *of the finite* p*-group* G. *Then, if* $k < p-1$,

$$s_k(i_!\rho) = i_*(s_k\rho) = i_*((c_1\rho)^k).$$

Since most of our calculations using Theorem 6.3 – notably for metacyclic p-groups, extra-special p-groups and non-abelian groups of order p^3 – only use this special case, we first give an argument for it using no more than the methods we have developed. Clearly we may also take p to be odd. Under the restrictive assumptions the flat bundle defined by the induced representation $i_!\rho$ is classified by the composition

$$\begin{array}{ccccccc}
BG & \longrightarrow & B(K \wr S_s) & \xrightarrow{B(\rho \wr 1)} & B(U(1) \wr S_s) & \longrightarrow & BU(s) \\
\uparrow & & \uparrow & & & \nearrow & \\
BK & \longrightarrow & B(\underbrace{K \times \cdots \times K}_{s}) & \longrightarrow & B(T^s) & &
\end{array}$$

where s denotes the index of K in G. The Newton polynomial $s_k(i_!\rho)$ is obtaining by restricting $s_k(\rho \wr 1) \in H^{2k}(K \wr S_s, \mathbb{Z})$ to the subgroup G, and in principle $s_k(\rho \wr 1)$ may be calculated using the spectral sequence of the extension defining the Wreath product. However with integral coefficients this does not collapse, and we can only say that $s_k(\rho \wr 1)$ equals the sum of a leading term from $H^0(S_s, H^{2k}(K \times \cdots \times K, \mathbb{Z}))$, which clearly restricts to $i_*((c_1\rho)^k)$ in the cohomology of G, and a number of correction terms belonging to subquotients of $H^{2i}(S_s, H^{2j}(K \times \cdots \times K, \mathbb{Z}))$, $i+j=k$. These correction terms all depend on the Chern classes $\{c_i(\pi_t); 1 \leqslant i \leqslant k, 1 \leqslant t \leqslant s\}$, where as in Chapter 7 π_t denotes the representation of the symmetric group S_t by permutation matrices. Hence each of these correction terms has order dividing the lowest common multiple of the denominators of B_i/i, i running though the even numbers less than k. Note that this step does not assume Theorem 6.3; den(B_i/i) gives an upper bound for the ith Chern class of any rational representation, and arises as a consequence of Galois invariance, see Chapter 7. By Von Staudt's theorem p does not divide den(B_i/i) for $i<p-1$, and so, in this range of dimensions, if G is a p-group the correction terms all vanish. For a more complete version of this argument, see [Th3]. We now return to the main theorem, i.e. we drop the restrictions that $k<p-1$, G is a p-group and ρ a one-dimensional representation of a normal subgroup K. Here are the main steps of the promised topological argument.

Generalised cohomology. We first need to introduce the cohomology theory $E^*(X)$ defined by the Ω-spectrum $\mathbf{E}$. For simplicity we suppose that $\mathbf{E} = \{E_0, E_1, E_2, \ldots\}$ is a sequence of topological spaces together with homotopy equivalences between E_{i-1} and the space of loops in E_i, $\Omega(E_i)$. For the present argument take $\mathbf{E} = \{\ldots, K(\mathbb{Z}, n), K(\mathbb{Z}, n+1), \ldots\}$, where $K(\mathbb{Z}, n)$ has trivial homotopy groups except for $\pi_n K \cong \mathbb{Z}$, and $\mathbf{F} = \mathbf{bu} = \{\ldots, \mathrm{BU}(2q, \ldots, \infty), \ldots\}$, where $\pi_i(\mathrm{BU}(2q, \ldots, \infty)) = 0$, $i<2q$, and $\pi_i(\mathrm{BU}(2q, \ldots, \infty)) = \pi_i \mathrm{BU}$ otherwise. Here BU is the classifying space defined in the last section of Chapter 5, and $\mathrm{BU}(2q, \ldots, \infty)$ classifies (stable) equivalence classes of bundles over $(2q-1)$-connected CW-complexes. Using the space of paths one easily sees that $\Omega K(\mathbb{Z}, n) \simeq K(\mathbb{Z}, n-1)$, and that **bu** is an evenly graded spectrum, that is $\Omega^2 \mathrm{BU}(2q, \ldots, \infty) \simeq \mathrm{BU}(2q-2, \ldots, \infty)$ as a consequence of the Bott periodicity isomorphism, see [Hs] Chapter 10.

Make the general definition $E^n(X) = [X, E_n]$, so that in particular

$$H^n(X, \mathbb{Z}) = [X, K(\mathbb{Z}, n)] \quad \text{and} \quad k^{2q}(X) = [X, \mathrm{BU}(2q, \ldots, \infty)].$$

The abstract transfer. Let $f: X \to Y$ be a finite covering map between finite dimensional CW-complexes, and consider pairs $(X, A = f^{-1}B)$, (Y, B). For a suitably large value of m we may embed $X \times I^m$ in $Y \times I^m$ so that the following diagram commutes up to homotopy:

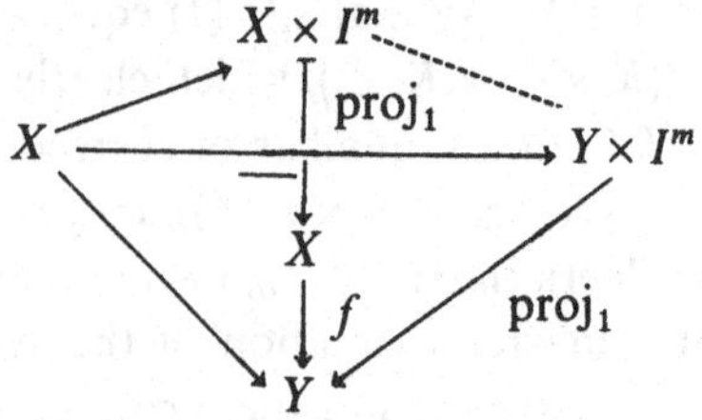

By mapping the complement of $X \times I^m$ to a point we may define

$$\begin{array}{ccc} \dfrac{Y \times I^m}{(Y \times \partial I^m) \cup (B \times I^m)} & \xrightarrow{f^{\dagger}} & \dfrac{X \times I^m}{X \times \partial I^m \cup (A \times I^m)} \\ \| & & \| \\ \Sigma^m(Y/B) & & \Sigma^m(X/A), \end{array}$$

where Σ denotes suspension. Define the transfer map in dimension n by

$$\begin{array}{ccc} [\Sigma^m(Y/B), E_{m+n}] & \xrightarrow{f^{\dagger *}} & [\Sigma^m(X/A), E_{m+n}] \\ \wr\| & & \wr\| \\ E^n(Y, B) & \xleftarrow[f_* = \mathrm{tr}_E(f)]{} & E^n(X, A) \end{array}$$

using the Σ–Ω duality in homotopy theory. If X and Y are infinite-dimensional, for example the classifying spaces BK and BG, we work with finite-dimensional skeleta X^k, and restrict attention to cohomology theories E^* such that $E^*(X) = \varprojlim_k E^*(X^k)$. This class includes $H^*(\cdot, \mathbb{Z})$ and k^*.

Cohomology operations For us a cohomology operation is a natural transformation of functors defined by a map of spectra $\mathbf{E} \to \mathbf{F}$; that is a sequence of maps $\lambda_{\cdot,r}\colon E_n \to F_{n+r}$ such that the diagram below commutes up to homotopy

$$\begin{array}{ccc} \Omega E_n & \xrightarrow{\Omega\lambda_{n,r}} & \Omega F_{n+r} \\ \wr\| & & \wr\| \\ E_{n-1} & \xrightarrow{\lambda_{n-1,r}} & F_{n+r-1} \end{array}$$

Such a cohomology operation commutes with the transfers tr_E and tr_F defined above, i.e. $\lambda_r \mathrm{tr}_E = \mathrm{tr}_F \lambda_r\colon E^n(X) \to F^{n+r}(Y)$ for all finite coverings. As an example we have the *Chern character*:

Let ξ be a complex vector bundle over the CW-complex X, and write

$$\mathrm{ch}_0\xi = \dim_{\mathbb{C}}\xi, \quad \mathrm{ch}_n\xi = \frac{s_n(c_1(\xi)\ldots c_n(\xi))}{n!} \in H^{2n}(X, \mathbb{Q}), \quad n \geqslant 1.$$

Collecting all the components together we have

$$\mathrm{ch}\colon K(X) = k^0(X) \longrightarrow \prod_{n\geqslant 0} H^{2n}(X, \mathbb{Q}), \text{ satisfying}$$

(i) if η is a line bundle, then $\mathrm{ch}(\eta) = e^{c_1(\eta)}$,
(ii) $\mathrm{ch}(\xi_1 \oplus \xi_2) = \mathrm{ch}(\xi_1) + \mathrm{ch}(\xi_2)$, and
(iii) $\mathrm{ch}(\xi_1 \otimes \xi_2) = \mathrm{ch}(\xi_1)\cdot\mathrm{ch}(\xi_2)$.

Since ch is natural, the splitting principle, Proposition 5.8, shows that (i)–(iii) determine the Chern character completely. Thus at the 0-level of the k^*-spectrum we have a family of maps

$$\mathrm{ch}_r\colon BU \to K(\mathbb{Q}, 2r),$$

which we wish to extend to higher levels by suspension. However we also wish to replace the coefficients $\mathbb{Q}$ in ordinary cohomology by $\mathbb{Z}$; thus we look for a family of maps

$$\mathrm{ch}_{q,r}\colon BU(2q, \ldots, \infty) \to K(Z, 2q + r)$$

such that $\Omega^2\mathrm{ch}_{q,r} \simeq \mathrm{ch}_{q-1,r}$ *and* such that in degree zero $\mathrm{ch}_{0,r}$ is rationally equivalent to $m(r)\mathrm{ch}_r$ for some suitable multiple $m(r)$. For a fixed value of r, $\mathrm{ch}_{q,r}\colon k^{2q}(X) \to H^{2q+2r}(X, \mathbb{Z})$ will then be a natural transformation of cohomology theories commuting with transfer. Setting $q = 0$ we obtain the central square in the diagram below

$$\begin{array}{ccccc}
R(K) & \longrightarrow & k^0(BK) & \xrightarrow{\mathrm{ch}_{0,r} = m(r)\mathrm{ch}_r} & H^{2r}(BK, \mathbb{Z}) = H^{2r}(K, \mathbb{Z}) \\
\downarrow i_! & & i_! \downarrow \mathrm{tr}_K & & \mathrm{tr}_K \downarrow i_* \qquad \downarrow i_* \\
R(G) & \longrightarrow & k^0(BG) & \xrightarrow{\mathrm{ch}_{0,r} = m(r)\mathrm{ch}_r} & H^{2r}(BG, \mathbb{Z}) = H^{2r}(G, \mathbb{Z}),
\end{array}$$

provided that we can show that the various definitions of transfer coincide. This can be done directly for the two theories concerned (for H^{2r} this is straightforward, for k^0 rather less so), or we can appeal to the following uniqueness result (L.G. Lewis, Uniqueness of bundle transfer, *Math. Proc. Camb. Phil. Soc.* **93** (1983), 87–111). Let $E^*(\cdot)$ be a cohomology theory such that if X is not finite dimensional, then $E^*(X) = \varprojlim_k E^*(X^k)$, that is $\lim^1 E^{*-1}(X) = 0$. Then any transfer map tr'_E coincides with the abstract transfer tr_E defined above for finite covering maps $f\colon X \to Y$, provided that

(i) tr'_E is natural with respect to pull-backs,
(ii) tr'_E is stable, i.e. defines a morphism between two long exact sequences for pairs, and
(iii) any component of Y in X contributes the identity.

The Adams construction We introduce the following notation: $j_*^{(0)}$ and $j_*^{(p)}$ are the maps induced in cohomology by the extension of scalars $\mathbb{Z}\to\mathbb{Q}$ and $\mathbb{Z}\to\mathbb{F}_p$, and $n_{q,q'}\colon BU(2q,\ldots,\infty)\to BU(2q',\ldots,\infty)$ is the natural map for $q\geqslant q'$. Define $m(r)=\prod_{p\leqslant 1+r}p^{[r/p-1]}$, so that $M_r\cdot r!=m(r)$, and $m(r)/m(r-1)=$ product of all primes p such that $p-1\,|\,r$. Then the central square of the diagram above commutes provided that there exist classes $c_{q,r}\in H^{2q+2r}(BU(2q,\ldots,\infty),\mathbb{Z})$ such that

(i) $\Omega^2\mathrm{ch}_{q,r}$ corresponds to $\mathrm{ch}_{q-1,r}$ under the Bott map $\Omega^2BU(2q,\ldots,\infty)\simeq BU(2q-2,\ldots,\infty)$, at least when $r-1+q>0$,
(ii) $j_*^{(0)}\mathrm{ch}_{q,r}=m(r)n_{q,1}^*\mathrm{ch}_{q+r}$, where ch_{q+r} is the $q+r=n$th component of the rational Chern character of the universal bundle.

The inductive construction of the classes $\mathrm{ch}_{q,r}$ uses the further conditions

(iii) $n_{q,q'}^*\mathrm{ch}_{q',r'}=m(r')/m(r)\mathrm{ch}_{q,r}$, $q\geqslant q'$, $q+r=q'+r'$, and
(iv) if $r=s(p-1)+t$ with s, t integral and $0\leqslant t<p-1$, then

$$j_*^{(p)}\mathrm{ch}_{q,r}=\begin{cases}(m(r)/p^s m(t))\chi(P^s)j_*^{(p)}\mathrm{ch}_{q,t}, & q+t>0\\ 0, & q+t=0.\end{cases}$$

Here $\chi(P^s)$ denotes the Steenrod reduced power twisted by the canonical antiautomorphism of A_p, the mod p Steenrod algebra. The suspension condition (i) shows that it is enough to verify (ii), (iii) and (iv) when q is large. But now an easy calculation in the rational cohomology spectral sequence associated to the fibration $n_{q,q'}$ shows that (ii) and (iii) may be replaced by

(ii)′ $j_*^{(0)}\mathrm{ch}_{q,0}=n_{q,1}^*\mathrm{ch}_q$, and
(iii)′ $n_{q+1,q}^*\mathrm{ch}_{q,r}=m(r)/m(r-1)\cdot\mathrm{ch}_{q+1,r-1}$.

However Bott's theory implies that the class $n_{q,1}^*\mathrm{ch}_q$, which is the leading term of the Chern character for bundles over spaces which are $(2q-1)$-connected is actually integral. Therefore $\mathrm{ch}_{q,0}$ exists. We calculate $\mathrm{ch}_{q,r}$ $(r\geqslant 1)$ using the cohomology spectral sequence of the fibration

$$\begin{array}{ccc}BU(2q+2,\ldots,\infty)\to & BU(2q,\ldots,\infty)\\ & \downarrow\\ & K(\mathbb{Z},2q)\end{array}$$

with coefficients in $\mathbb{Z}$ and $\mathbb{F}_p$. Assume that the class $\mathrm{ch}_{q+1,r-1}$ has been constructed. Then for $q \gg r$ the image under transgression $\tau\mathrm{ch}_{q+1,r-1}$ has finite order in $H^{2q+2r+1}(K(\mathbb{Z},2q),\mathbb{Z})$; we now reduce modulo p and determine this order. It turns out that this is where condition (iv) for the pair $(q+1, r-1)$ enters the argument, that $\tau\mathrm{ch}_{q+1,r-1}$ has order equal to the product of the distinct primes p such that $(p-1)$ divides r, i.e. $m(r)/m(r-1)$. Hence we can choose *some* class $e \in H^{2q+2r}(BU(2q,\ldots,\infty),\mathbb{Z})$ such that $n^*_{q+1,q}e = m(r)/m(r-1)\cdot\mathrm{ch}_{q+1,r-1}$. The class $\mathrm{ch}_{q,r}$ is then taken to be that e which ensures that condition (iv) extends to the pair (q,r). However the argument is highly technical, and we can do no more than refer the reader to [Ad1] for full details.

Appendix 2

Integral cohomology of non-abelian groups of order p^3, $p \geqslant 3$

Type I:
This is a special case of Theorem 4.4.

Type II:
$G = \{A, B, C: A^p = B^p = C^p = [A, C] = [B, C] = 1, \quad [A, B] = C\}$.
Let $\hat{\alpha}$ and $\hat{\beta}$ be 1-dimensional representations of G mapping A and B to primitive pth roots of unity, and as usual write $\alpha = c_1(\hat{\alpha})$, $\beta = c_1(\hat{\beta})$. Let γ be the 1-dimensional representation of the subgroup $\langle B, C\rangle$ which maps C to a primitive pth root and B to 1.

Then we have the following table

Generator	α	β	$\chi_j = i_* \gamma^j$	$\xi = c_p(i_! \hat{\gamma})$	μ	ν
Additive order	p	p	p	p^2	p	p
Dimension	2	2	$2 \leqslant j \leqslant p-1$	$2p$	3	3

Relations:

(1) $\alpha\mu = \beta\nu$,
(2) $\alpha^p\mu = \beta^p\nu$,
(3) $\mu^2 = \nu^2 = 0$,
(4) $\chi_j\chi_k = \alpha\chi_j = \beta\chi_j = \mu\chi_j = \nu\chi_j = 0,\ 2 \leqslant j,\ k \leqslant p-1$,
(5) $\chi_j\chi_{p-1} = 0,\ 2 \leqslant j < p-1,\ \chi_{p-1}^2 = \alpha^{p-1}\beta^{p-1}$,
(6) $\alpha\chi_{p-1} = \alpha\beta^{p-1},\ \beta\chi_{p-1} = \beta\alpha^{p-1}$,
(7) $\mu\chi_{p-1} = \mu\alpha^{p-1},\ \nu\chi_{p-1} = \nu\beta^{p-1}$,
(8) $\alpha\beta^p = \beta\alpha^p$.

If $p > 3$ then $\chi_3 = d\mu\nu$ for some unit $d \in \mathbb{Z}/p$. If $p = 3$ then $p\xi = e\mu\nu$ with $e = \pm 1$.

If t_1 is the automorphism which interchanges A and B, and t_2 is the automorphism which fixes B and maps A to AB, then under the induced automorphisms in cohomology we have

(i) $\alpha \longmapsto \beta$, $\mu \longmapsto -\nu$, $\chi_j \longmapsto \pm\chi_j$, and

(ii) $\alpha \longmapsto \alpha$, $\beta \longmapsto \beta+\alpha$, $\nu \longmapsto \mu+\nu$, $\chi_j \longmapsto \chi_j$ $(2 \leqslant j < p-1)$,
$\chi_{p-1} \longmapsto \chi_{p-1} + (\beta+\alpha)^{p-1} - \beta^{p-1}$

Remarks. With minor changes, in notation the information above is taken from [Le], Theorem 6.26. As should be clear from the partial calculations in Chapter 8, the two odd dimensional generators are detected by $E_2^{1,2}$ in the spectral sequence of the central extension of C_p^C by $\langle \bar{A}, \bar{B} \rangle$. The multiplicative relations (4) between the even-dimensional generators may be proved by first using the double coset formula, Lemma 3.1(ii), to show that $i^*\chi_j = 0$ for *all* proper subgroups of G and then applying Frobenius reciprocity, Lemma 2.6(ii). The anomalous behaviour of χ_{p-1} in (5), (6) and (7) compared to that of χ_j $(2 \leqslant j \leqslant p-2)$ can be explained in terms of the Riemann–Roch formula. For all except the top values of j the generator χ_j equals the Newton polynomials $s_j(i, \hat{\gamma})$, but $\chi_{p-1} = s_{p-1}(i, \hat{\gamma}) +$ a correction term, see Lemma 8.7. With the exception of the odd dimensional generators μ, ν (for which inspection of the spectral sequence is again necessary) the effect of the automorphisms t_1 and t_2 may be read off from the effect on the representations of G and its subgroup $\langle B, C \rangle$. Some of these observations serve to clarify parts of the argument in [Le], and again illustrate the usefulness of Chern classes in calculations.

Appendix 3

Non-abelian groups of order p^4, $p \geqslant 5$

The list below is taken from W. Burnside, *Theory of finite groups*, Cambridge University Press (1897), pages 87–8. A less explicit reference is [Hp, II, Satz 12.6], where the classification is obtained in terms of Blackburn's Theorem 8.5.

In each case the generators are taken from the four symbols A, B, C, D.

(1) $A^{p^3} = B^p = 1$, $A^B = A^{1+p^2}$, metacyclic.

(2) $A^{p^2} = B^p = C^p = [A,B] = [A,C] = 1$, $[B,C] = A^p$.

(3) $A^{p^2} = B^{p^2} = 1$, $A^B = A^{1+p}$, metacyclic.

(4) $A^{p^2} = B^p = C^p = [A,B] = [B,C] = 1$; $A^C = A^{1+p}$, the group is isomorphic to $C_p^B \times P_1^{\langle A,B \rangle}$.

(5) $A^{p^2} = B^p = C^p = 1$, $[A,C] = B$, $[A,B] = [B,C] = 1$.

(6–8) $A^{p^2} = B^p = [B,C] = 1$, $[A,C] = B$, $A^B = A^{1+p}$, $C^p = A^{sp}$, where $s = 0$, 1 or a quadratic non-residue modulo p.

(9) $C_p^B \times P_2^{\langle A,C \rangle}$, compare (4) above.

(10) The group has four generators and exponent p:

$$[A,B] = [A,C] = [A,D] = [B,C] = 1, \quad [C,D] = B, \quad [B,D] = A.$$

(1), (2), (3), (7) and (8) have p-rank equal to two, see Theorem 8.5 with a change of presentation for the third type. Group (5) is a minimal non-abelian group whose cohomology can be calculated using a spectral sequence argument similar to that for P_2 of order p^3. In principle $H^{\text{even}}(C_p \times P_i)$, $i = 1, 2$, can be calculated from the results in Appendix 2 plus the Künneth formula. This leaves groups (6) and (10) as the next examples to be considered.

References

Books

Ba A Babakhanian, *Cohomological methods in group theory*, Marcel Dekker (New York), 1972

Br K. Brown, *Cohomology of groups*, Springer Verlag (Heidelberg), 1982

B-Sh Z. Borevich–I. Shafarevich, *Number Theory*, Academic Press, 1966

C-E H. Cartan–S. Eilenberg, *Homological Algebra*, Princeton University Press, 1956

Gb K.W. Gruenberg, *Cohomological topics in group theory*, LN143, Springer-Verlag (Heidelberg), 1970

Hp B. Huppert, *Endliche Gruppen I*, Springer-Verlag (Heidelberg), 1966

Hs D. Husemoller, *Fibre bundles*, First edition, McGraw-Hill, 1966

La S. Lang, *Cohomologie des groupes*, Benjamin, (New York) 1966

Mac S. Maclane, *Homology*, Springer-Verlag (Heidelberg), 1963

Mi J. Milnor, *Characteristic classes*, Annals of Math. Studies 76, 1974, Princeton University Press

NB N. Bourbaki, *Lie groups and algebras*, Chapters II–III, Herman/Addison-Wesley, 1972

Se1 J.-P. Serre, *Corps locaux*, Hermann (Paris), 1968

Se2 J.-P. Serre, *Représentations linéaires des groups finis*, Herman (Paris), 1967

Sp E. Spanier, *Algebraic topology*, McGraw-Hill, 1966

Wo J. Wolf, Spaces of constant curvature, various editions, McGraw-Hill/Publish or Perish.

Articles

Ad1 J.F. Adams, On Chern characters and the structure of the unitary group. *Proc. Camb. Phil. Soc.* **57**(1961) 189–95

Ad2 J.F. Adams, Stable homotopy theory, LN2 Springer-Verlag, 1964

A1 K. Alzubaidy, Thesis, London University, 1979.

Ar D. Arletaz, Chern–Klassen von ganzzahligen und rationalen Darstellungen diskreter Gruppen, *Math. Z.* **187**(1984), 49–60

At M.F. Atiyah, Characters and the cohomology of finite groups, *Publ. Math. IHES* **9**(1961) 23–64

A-T M.F. Atiyah–D. Tall, Group representations, λ-rings and the *J*-homomorphism, *Topology* **8** (1969)253–97.

Bl N. Blackburn, Generalisations of certain elementary theorems on *p*-groups. *Proc. London Math. Soc.* **11**(1961) 1–22

Bo E.H. Boorman, *S*-operations in representation theory, *Trans. Amer. Math. Soc.* **205**(1975) 127–49

Ch R. Charney, Homology stability of GL_n of a Dedekind domain, *Bull. Amer. Math. Soc.* (NS) **1**(1979) 428–31

C-V L. Charlap–A. Vasquez, The cohomology of group extensions, *Trans. Amer. Math. Soc.* **124**(1966) 24–40

Ev1 L. Evens, The cohomology ring of a finite group, *Trans. Amer. Math. Soc.* **101**(1961) 224–39

Ev2 L. Evens, A generalisation of the transfer map in the cohomology of groups, *Trans. Amer. Math. Soc.* **108**(1963) 54–65

Ev3 L. Evens, On the Chern classes of representations of finite groups, *Trans. Amer. Math. Soc.* **115**(1965) 180–93

E-K1 L. Evens–D. Kahn, Chern classes of certain representations of symmetric groups, *Trans. Amer. Math. Soc.* **245**(1978/9) 309–30

E-K2 L. Evens–D. Kahn, An integral Riemann–Roch formula for induced representations of finite groups, *Trans. Amer. Math. Soc.* **245**(1978/9) 331–47

E-M B. Eckmann–G. Mislin, Chern classes of group representations over a number field, *Comp. Math.* **44**(1981) 41–65

F T. Farrell, An extension of Tate cohomology to a class of infinite groups *J. Pure and Applied Alg* **10**(1977) 153–61

G A. Grothendieck, Classes de Chern et représentations linéaires des groupes discrets, *10 exposés sur la cohomologie étale des schémas*, North-Holland 1968, 215–305

Gr J.A. Green, On the characters of the finite general linear group, *Trans. Amer. Math. Soc.* **80**(1955) 402–47

J J.P. Jouanolou, Exposé VII SGA 5, Cohomologie *l*-adique et fonctions *L*, LN 589, Springer-Verlag (1977) 282–350

J-Z S. Jackowski–T. Zukowski, *P*-free linear representations of *P*-solvable finite groups, LN 763, Springer-Verlag (1979) 458–64

Kr P. Kropholler, to appear

Ku T.N. Kuo, On the exponent of $H^n(G, \mathbb{Z})$, *J. of Algebra* **7**(1967) 160–7

Le G. Lewis, Integral cohomology rings of groups of order p^3, *Trans. Amer. Math. Soc.* **132**(1968) 501–29

Mn B. Mann, The cohomology of the symmetric groups, *Trans. Amer. Math. Soc.* **242**(1978) 157–84

N M. Nakaoka, Homology of the infinite symmetric group, *Annals of Math.* **73**(1961) 229-57

P-Sh I. Piatetskii–Shapiro, Complex representations of SL(2,K) for finite fields K, *Contemporary Math.* **16**(1983), 71 pages

Q1 D. Quillen, The Adams Conjecture, *Topology* **10**(1970) 67–80

Q2 D. Quillen, The spectrum of an equivariant cohomolgy ring I, *Annals of Math.* **94**(1971) 549–72

Q3 D. Quillen, The mod 2 cohomology rings of extra-special 2-groups and the spinor groups, *Math. Ann.* **194**(1971) 197–212

Q4 D. Quillen, On the cohomology and K-theory of the general linear groups over a finite field, *Annals of Math.* **96**(172) 552–86

Q5 D. Quillen, Text of a letter to J. Milnor, *Algebraic K-theory Proceedings*, LN551, Springer-Verlag (1976) 182–8

Sc I. Schur, Über die Darstelling der endliche Gruppen durch gebrochene lineare Substitutionen, *J. Reine u.a. Mathematik* **127**(1904) 20–50

Sg G. Segal, Equivariant K-theory, *Publ. Math. IHES* **34**(1968) 129–51

So C. Soulé, K-théorie des anneaux d'entiers de corps de membres et cohomologie étale, *Inv. Math.* **55**(1979) 251–95

St U. Staffelbach, Thesis, ETH Zürich, 1983

Sw1 R.G. Swan, Induced representations and projective modules, *Annals of Math.* **71**(1960) 552–78

Sw2 R.G. Swan, The p-period of a finite group, *Ill. J. Math.* **4**(1960) 341–6

Th1 C.B. Thomas, Chern classes and groups with periodic cohomology, *Math. Ann.* **190**(1971) 323–8

Th2 C.B. Thomas, Riemann–Roch formulae for group representations, *Mathematika* **20**(1973) 253–62

Th3 C.B. Thomas, An integral Riemann–Roch formula for flat line bundles, *Proc. London Math. Soc.* (3) **34**(1977) 87–101

Th4 C.B. Thomas, On Poincaré 3-complexes with binary polyhedral fundamental group, *Math. Annalen* **226**(1977) 207–21

Th5 C.B. Thomas, Characteristic classes of representations over imaginary quadratic fields, LN788 Springer-Verlag (1980) 471–81

Th6 C.B. Thomas, Admissible filtrations on the representation ring of a finite group, *Contemporary Mathematics* **19**(1983)

Wa C.T.C. Wall, Resolutions for extensions of groups, *Proc. Camb. Phil. Soc.* **57**(1961) 251–5

Ya N. Yagita, to appear.

Index of symbols

Index